Automated Mathematical Induction

Automated Mathematical Induction

Edited by

H. ZHANG

University of Iowa, USA

Reprinted from *Journal of Automated Reasoning* 16 (Nos. 1–2): 1–222, 1996

Kluwer Academic Publishers

Dordrecht / Boston / London

Library of Congress Cataloging-in-Publication Data

A C.I.P. Catalogue record for this book is available from the Library of Congress.

ISBN 0-7923-4010-8

Published by Kluwer Academic Publishers,
P.O. Box 17, 3300 AA Dordrecht, The Netherlands.

Kluwer Academic Publishers incorporates
the publishing programmes of
D. Reidel, Martinus Nijhoff, Dr W. Junk and MTP Press.

Sold and distributed in the U.S.A. and Canada
by Kluwer Academic Publishers,
101 Philip Drive, Norwell, MA 02061, U.S.A.

In all other countries, sold and distributed
by Kluwer Academic Publishers Group,
P.O. Box 322, 3300 AH Dordrecht, The Netherlands.

Printed on acid-free paper

Printed in the Netherlands

Table of Contents

Preface to the Special Issue on Automated Mathematical Induction

Two decades ago, Boyer and Moore built one of the first automated theorem provers that are capable of proofs by mathematical induction. Today, the Boyer–Moore theorem prover remains the most successful in the field. For a long time, the research on automated mathematical induction was confined to very few people. In recent years, as more people realize the importance of automated inductive reasoning to the use of formal methods of software and hardware development, more automated inductive proof systems have been built.

Three years ago, the interested researchers in the field formed two consortia on automated inductive reasoning – the MInd consortium in Europe and the IndUS consortium in the United States. The two consortia organized three joint workshops in 1992–1995. There will be another one in 1996. Following the suggestions of Alan Bundy and Deepak Kapur, we edited this special issue to document advances in understanding of the field and in the power of the theorem provers that can be built.

The call for papers for the special issue attracted many interesting submissions, and each of them was carefully refereed by the experts in the field. We are very grateful both to those who submitted papers and to the anonymous referees.

Six papers were selected for this issue. Unlike other special issues of this journal, we provide the reader with a tutorial study of the Boyer–Moore theorem prover. The other five papers present novel ideas that could be used to build theorem provers more powerful than the Boyer–Moore theorem prover.

HANTAO ZHANG
University of Iowa
Iowa City
IA 52242, U.S.A.

e-mail: hzhang@cs.uiowa.edu

Journal of Automated Reasoning **16:** 3–37, 1996.

Induction Using Term Orders*

FRANCOIS BRONSARD**
CRIN-INRIA-Lorraine, Nancy, France; Current address: Andersen Consulting, 3773 Willow road, #2f5d, Northbrook, IL 60062-6212, U.S.A.

UDAY S. REDDY
The University of Illinois at Urbana-Champaign, U.S.A., e-mail: reddy@cs.uiuc.edu

ROBERT W. HASKER
The University of Illinois at Urbana-Champaign, U.S.A., e-mail: hasker@cs.uiuc.edu

Abstract. We present a procedure for proving inductive theorems which is based on explicit induction, yet supports *mutual induction*. Mutual induction allows the postulation of lemmas whose proofs use the theorems *ex hypothesi* while the theorems themselves use the lemmas. This feature has always been supported by induction procedures based on Knuth–Bendix completion, but these procedures are limited by the use of rewriting (or rewriting-like) inferences. Our procedure avoids this limitation by making explicit the implicit induction realized by these procedures. As a result, arbitrary deduction mechanisms can be used while still allowing mutual induction.

Key words: Mutual induction, automated theorem proving.

1. Introduction

Existing proof methods for automated inductive theorem proving fall into two camps: *explicit induction* using induction schemes, as implemented for example in Nqthm [5, 6] and other systems [3, 11, 38], and *implicit induction* based on deduction procedures like Knuth–Bendix completion [1, 12, 15–17, 20–22, 25, 27–31]. Both the methods have professed advantages. Explicit induction has the flexibility of allowing arbitrary well-founded orders (or termination functions) for doing induction. Implicit induction has the benefit of allowing 'mutual induction', where a theorem and a lemma can appeal to each other in their proofs. In this paper, we formulate a proof method that combines the benefits of both these methods. It is based on explicit induction using *propositional orders* – well-founded orders on terms representing propositions. This method retains the benefits of mutual induction available in the implicit method. At the same time, it allows arbitrary well-founded orders to be incorporated in the propositional order whereas the implicit methods are limited to the order implicit in the deduction mechanism.

* A preliminary version of this paper appeared in the proceedings of the *12th Conference on Automated Deduction*, A. Bundy, editor.

** This author was supported by a grant from the Ministère des Affaires Etrangères, France. Current address: Andersen Consulting, Chicago, U.S.A.; e-mail: bronsard@andersen.com.

'Mutual induction' is the phenomenon where a theorem is proved using a lemma in such a way that both the theorem and the lemma are used in each other's proofs. (The lemma is obviously needed in the proof of the theorem. But, the proof of the lemma can also appeal to the theorem *ex hypothesi*.) Such proofs clearly arise in proving properties of mutually recursive function definitions. However, it has been noticed that this phenomenon occurs much more often in automated proof systems based on implicit induction. In order to prove a theorem P, the system would find a collection of subgoals $P_1, \ldots, P_n$. Some of these goals would be directly provable, but others reduce to simpler formulas $Q_1, \ldots, Q_k$ that need further inductive proof. It so happens that the system often needs to appeal to the theorem P in proving the subgoals $Q_1, \ldots, Q_k$. As noted in [31], this phenomenon arises in proving something as simple as the commutativity of addition. Mutual induction is much less likely to arise in hand-crafted proofs where we would be careful to formulate lemmas that can be proved independently. But, in an automated system where 'insight' is in short supply, we would like to use any formula that the system encounters as a potential hypothesis, and such hypotheses are likely to be interdependent. When mutual induction is freely available, all this happens effortlessly, leading to a high degree of automation in the proof. For example, this phenomenon leads to a dramatic reduction in the number of user-supplied lemmas in proving the Gilbreath card trick theorem [4].

Implicit induction methods support mutual induction as they rely on a global term order that is implicit in the deduction mechanism. When the deduction mechanism is *reductive*, i.e., always reduces goal propositions by the global term order, it is possible to guarantee that all uses of the inductive hypothesis are applied to instances of the theorem smaller than the one being proved. Thus, the soundness of mutual induction crucially depends on the reductive mechanisms used for deduction.

Unfortunately, this very fact makes the implicit induction methods (relatively) incomplete [31]. When reductive mechanisms are used for first-order (i.e., non-inductive) reasoning, they are used in conjunction with powerful *forward* deduction mechanisms (such as superposition, ordered resolution, oriented paramodulation etc.). Reductive mechanisms, by themselves, are incomplete for first-order reasoning. Since implicit induction methods rely on reductive mechanisms for their soundness, it is not possible to incorporate forward deduction mechanisms. So, the first-order reasoning components of these methods are necessarily incomplete.

Our solution to this problem is to do induction with explicit reference to a global propositional order. Because the order is made explicit, we do not need to appeal to any implicit order present in the deduction mechanisms. Second, we devise methods to track the dependencies on induction hypotheses through forward deduction inferences. Thus, the first-order component of the reasoning is freed from the limitations of reductive mechanisms and complete semi-decision procedures can be incorporated. At the same time, the benefits of mutual induction are retained because the induction is with respect to a global proposition order. In fact, every implicit induction proof is also a valid proof using our method. So, all the mutual induction power of implicit induction is still available.

Another important aspect of our treatment in this paper is that we take an axiomatic approach to first-order deduction systems. This not only leads to generality, covering a wide class of deduction mechanisms such as rewriting, conditional rewriting and resolution-based methods for clauses, but also leads to a high degree of abstraction where the essential content of the induction methods can be seen clearly. We feel that the large body of work that exists on implicit induction intermixes the first-order deduction aspects (mainly rewriting) with the induction aspects a little too much. Abstraction of the first-order deduction aspects using an axiomatization is highly rewarding, almost to the point of making our results on induction look entirely obvious.

This paper is a successor to [31], where implicit induction was formulated for the term rewriting deduction mechanism. In this paper, we formulate both implicit and explicit induction for a wide class of deduction mechanisms. We also show how to incorporate forward inferences, which is a problem left open in [31]. We know of no other work that incorporates forward inferences in the framework of implicit induction, except for a brief reference in an early version of [27] where it was incorrectly supposed that the result of forward inferences ('inessential critical pairs') could be used for simplification without any test. (This was later corrected.) A preliminary version of the present paper appeared as [10].

The organization of this paper is as follows. In Section 2, we review background material on propositional orders and inductive consequence. We also define an axiomatization of deductive systems with reference to which inductive proof procedures are later defined. In Sections 3 and 4, we define the basic induction method using propositional orders and show how it supports mutual induction. Sections 5 and 6 assume further structure on deduction systems in terms of backward and forward inferences, and adapt the inductive proof procedure to track dependencies in forward inferences. In Section 7, we show the relationship of our method to implicit induction and comment on its completeness.

2. Preliminaries

2.1. Deduction Systems

Because our inductive proof procedure is applicable to many kinds of deduction systems, we use a fairly general notion of deduction system:

DEFINITION. A (first-order) *deduction system* is specified by the following data:

- F, a finite, ranked alphabet of *function symbols*,
- P, a finite, ranked alphabet of *predicate symbols*,
- X, a countable set of variables,
- $\mathcal{L}$, a recursive set of *propositions* over F, P, and X, and
- $\vdash_V \subseteq (\mathbb{P}_{\text{rec}}\mathcal{L} \times \mathcal{L})$, a family of recursively enumerable *deduction* relations indexed by subsets $V \subseteq_{\text{fin}} X$ (where $\mathbb{P}_{\text{rec}}\mathcal{L}$ denotes recursive sets over $\mathcal{L}$.)

The deduction relation is subject to the following axioms:

(i) $\phi \in \mathcal{A} \Rightarrow \mathcal{A} \vdash_V \phi$.
(ii) $\mathcal{A} \vdash_V \phi \wedge \mathcal{A} \subseteq \mathcal{A}' \Rightarrow \mathcal{A}' \vdash_V \phi$.
(iii) $\mathcal{A} \vdash_V \phi \wedge \mathcal{A} \cup \{\phi\} \vdash_V \phi' \Rightarrow \mathcal{A} \vdash_V \phi'$.
(iv) $\mathcal{A} \vdash_V \phi \Rightarrow \mathcal{A} \vdash_V \phi\theta$ for all substitutions θ such that $\text{dom}(\theta) \cap V = \emptyset$.
(v) $\mathcal{A} \vdash_V \phi \Rightarrow \mathcal{A}\theta \vdash_{V'} \phi\theta$ for all substitutions θ mapping V to terms over $V' \subseteq_{\text{fin}} X$.

We write $\mathcal{A} \vdash \phi$ for $\mathcal{A} \vdash_\emptyset \phi$ and $\mathcal{A} \vdash^0 \phi$ for $\mathcal{A} \vdash_{FV(\phi)} \phi$, where FV denotes the set of free variables of ϕ.

The axiom (iv) implies that all variables outside of V are implicitly universally quantified in each proposition. We normally assume that axioms in $\mathcal{A}$ are renamed apart from each other and from V so as to avoid name conflicts. The variables of V are 'eigenvariables', i.e., variables that are treated as constants in a particular deduction. The axiom (v) implies that these variables can be globally replaced by arbitrary terms.

We use the following conventions: $\mathcal{T}$ (and $\mathcal{T}_G$) denotes terms (and ground terms) over F and X; $\mathcal{P}$ (and $\mathcal{P}_G$) denotes atoms (and ground atoms) over P, F, and X; and $\mathcal{L}_G$ denotes ground (or closed) propositions.

EXAMPLE 1. An *equational* deduction system has an empty set of predicate symbols. A proposition is a multiset $\{t, u\}$ of two terms, normally written as $t = u$ (or $u = t$). Let $\leftrightarrow_A$ be the relation defined by $t[a\theta/\alpha] \leftrightarrow_A t[b\theta/\alpha]$ whenever α is a position in t, $a = b$ is an equation in $\mathcal{A}$ and θ is a substitution. The deduction relation is

$$\mathcal{A} \vdash^0 t = u \quad \text{iff} \quad t \leftrightarrow^*_A u.$$

EXAMPLE 2. The deduction system of *conditional equations* also has an empty set of predicate symbols. A proposition is a pair $(\{e_1, \ldots, e_n\}, e)$ where $e_1, \ldots, e_n$ and e are equations. It is normally written as $e_1 \wedge \cdots \wedge e_n \Rightarrow e$. The deduction relation combines the standard equational axioms (reflexivity, symmetry, transitivity, congruence) and the cut (or modus ponens) inference [36].

EXAMPLE 3. A clausal deduction system has, as its propositions, multisets $\{l_1, \ldots, l_n\}$ of literals (i.e., signed atoms). We normally write such a proposition as $l_1 \vee \cdots \vee l_n$ (or as $l_1 \vee \cdots \vee l_k \Leftarrow \neg l_{k+1} \wedge \cdots \wedge \neg l_n$). The deduction relation is

$$\mathcal{A} \vdash^0 l_1 \vee \cdots \vee l_n \quad \text{iff} \quad \mathcal{A} \cup L \vdash_R \bot$$

where L is the skolemized negation of $l_1 \vee \cdots \vee l_n$ and $\vdash_R$ is the deduction mechanism built using the resolution and factoring inferences [34].

EXAMPLE 4. A clausal deduction system with *equality* has, among its predicate symbols, a distinguished symbol '$=$' of arity 2. The deduction relation is the clausal deduction mechanism augmented with the paramodulation inference (or the appropriate axioms such that '$=$' is an equivalence relation) [33, 26].

2.2. PROPOSITIONAL ORDERS

Briefly, we recall the basic notions. A *preorder* $\leqslant$ is a reflexive, transitive relation. The equivalence $\approx$ of a preorder is defined by $x \approx y \Leftrightarrow x \leqslant y \wedge y \leqslant x$. The *partial order* $<$ of $\leqslant$ is: $x < y \Leftrightarrow x \leqslant y \wedge x \not\approx y$. The reflexive closure of $<$ is denoted $\leq$, which is in general different from $\leqslant$. A partial order $<$ is said to be *well founded* if there is no infinite descending sequence $x_1 > x_2 > \cdots$. A preorder is said to be well founded if its partial order is well founded.

By a *term order*, we mean a well-founded order $\prec$ on the set of terms and atoms $(\mathcal{T} \cup \mathcal{P})$ that is stable, i.e., $t \prec u \Rightarrow t\theta \prec u\theta$ for all substitutions θ, and compatible with function and predicate symbols, i.e., $t \prec u \Rightarrow f(\ldots, t, \ldots) \prec f(\ldots, u, \ldots)$. A *propositional order* is a stable, well-founded preorder $\preccurlyeq$ on the set of proposition $\mathcal{L}$.

EXAMPLE 5 (Multiset path order). A classical term order is the *multiset path order* due to Dershowitz [14] (also called the recursive path order). Let $<_p$ be a partial order over the alphabet F of function symbols (called *precedence*). The *multiset path order* $\prec$ generated by $<_p$ is a partial order on $\mathcal{T}$ defined as follows:

$$s = f(s_1, \ldots, s_n) \prec g(t_1, \ldots, t_m) = t$$

if and only if any one of the following condition holds:

- $s \preceq t_i$ for some $i, 1 \leq i \leq m$, or
- $f <_p g$ and $s_j \prec t$ for all $j, 1 \leq j \leq n$, or
- $f = g$ and $\{s_1, \ldots, s_n\} \prec_{\text{mul}} \{t_1, \ldots, t_m\}$,

where $\prec_{\text{mul}}$ is the multiset extension of $\prec$ to multisets of terms. It is defined by

$$X \cup \{y_1, \ldots, y_n\} \prec_{\text{mul}} X \cup \{x\} \quad \text{if} \quad y_i < x \quad \text{for all } i, 1 \leq i \leq n.$$

Note that the multiset path order is stable and compatible with function symbols. In addition, it includes the subterm ordering, i.e., $s \prec t$ whenever s is a proper subterm of t.

To compare propositions (which are often made of multisets), we use an extension of term orders to multisets, called the *max-extension.*

DEFINITION (Max-extension). Let $<$ be a partial order on a set S, and let $\max(X)$ denote the multiset of $<$-maximal elements of the multiset X. The *max-extension* of $<$ is a preorder $\leqslant_{\max}$ on multisets over S defined by

$$X \leqslant_{\max} Y \Leftrightarrow \max(X) <_{\text{mul}} \max(Y) \vee \max(X) = \max(Y).$$

It may be verified that

$$X \approx_{\max} Y \overset{\text{def}}{\Longleftrightarrow} X \leqslant_{\max} Y \wedge Y \leqslant_{\max} X \Leftrightarrow \max(X) = \max(Y)$$
$$X <_{\max} Y \overset{\text{def}}{\Longleftrightarrow} X \leqslant_{\max} Y \wedge X \not\approx_{\max} Y \Leftrightarrow \max(X) <_{\text{mul}} \max(Y).$$

LEMMA 1. *If $<$ is a well-founded order, its max-extension is a well-founded preorder.*

EXAMPLE 6. We use the max-extension preorder to define propositional orders for the various deductive systems of Examples 1–4:

1. For the equational deduction system, let $\prec$ be a well-founded order on $\mathcal{T}$. Then $\preccurlyeq_{\max}$ can be used as a well-founded preorder on equations, i.e., $(t_1 = t_2) \preccurlyeq_{\max} (u_1 = u_2)$ iff $t_1 \preccurlyeq u_i$ and $t_2 \preccurlyeq u_j$ (for some $i, j \in \{1, 2\}$). For example, $(x + 0 = x) \prec_{\max} (S(x) + 0 = S(x))$ and $(x + 0 = x) \prec_{\max} (x{+}0 = 0{+}x)$ where $\prec$ is the multiset path order defined with the precedence $+ >_p S >_p 0$.
2. For clauses, let $\prec$ be a well-founded order on atoms $(\mathcal{P})$. We extend it to literals by ignoring their sign and use $\prec_{\max}$ as well-founded order on clauses.
3. For clauses with equality, let $\prec$ be a well-founded order on terms and non-equality atoms. The order $\prec_A$ on atoms is obtained by treating equality atoms $t_1 = t_2$ as binary multisets $\{t_1, t_2\}$ and other atoms a as unary multisets $\{a\}$, and comparing them by the max extension of $\prec$. The order on clauses is then $(\prec_A)_{\max}$.

Note that all these orders satisfy $t \prec u \Rightarrow \phi[t] \preccurlyeq \phi[u]$. However, $t \prec u$ does not necessarily imply $\phi[t] \prec \phi[u]$.

The need for propositional orders using max-extension arises from our interest in implicit induction procedures. As explained in Section 7, implicit induction procedures can be formalized as induction procedures using propositional orders. For existing implicit induction procedures this formalization relies on propositional orders using max-extension.

2.3. Induction

We take the following to be our operational definition of inductive consequence:

DEFINITION (Inductive consequence). Given a deduction system, let $\mathcal{A}$ be a set of axioms. A proposition ϕ is an *inductive consequence* of $\mathcal{A}$, written $\mathcal{A} \vdash_{\text{ind}} \phi$, if and only if, for every ground instance $\phi\alpha$ of ϕ, $\mathcal{A} \vdash \phi\alpha$.

The above definition is classical. (See [37] for a discussion.) From a model-theoretic point of view it defines inductive consequences as those properties that hold in all reachable models (or, equivalently, all term models). Sometimes, variants of this definition may be preferred. For example, in multi-sorted theories, one uses ground instances that respect the sorts. If some functions are only partially defined, it is more intuitive to restrict instantiations to constructor terms. In this case, the models of interest are the *constructor models*. In the context of a parameterized specification, we might allow instantiation only of variables of non-parameter sorts. This would allow models with arbitrary domains for the parameter sorts. The methods of this paper can be easily adapted to such variants.

In [31], Noetherian induction on ground terms was used to prove inductive consequences:

PROPOSITION 2 (Noetherian induction principle using term orders). *Let $\prec$ be a well-founded order over $\mathcal{T}_G$, $\mathcal{A}$ be a set of axioms, and $\phi[x]$ be a proposition. $\mathcal{A} \vdash_{\text{ind}} \phi[x]$ if, for all ground terms g,*

$$\mathcal{A} \cup \{\phi[g'] \mid g' \prec g\} \vdash \phi[g].$$

There is a qualitative difference between Noetherian induction using semantic orders and that using term orders. It is possible for $g' \prec g$ to be satisfied even if the two ground terms denote the same semantic value. For instance, $(n_1 + n_2) + n_3 \prec n_1 + (n_2 + n_3)$ holds in a lexicographic path order even though the two terms presumably have the same semantics. In a sense, the induction is carried out at the meta-level, and the definition of inductive consequence allows us to transfer the conclusions down to object-level reasoning.

In the rest of the paper, we extend this notion to an induction principle based on propositional orders, i.e., well-founded orders over $\mathcal{L}_G$ or $\mathcal{L}$. This extension allows us to incorporate forward inferences with relative ease.

3. Induction Using Propositional Orders

Given a well-founded order over ground propositions, i.e., an order over $\mathcal{L}_G$, the Noetherian induction principle using propositional orders can be expressed as follows:

PROPOSITION 3 (Noetherian induction using propositional orders). *Let $\preccurlyeq$ be a well-founded preorder over $\mathcal{L}_G$, $\mathcal{A}$ be a set of axioms, and $\phi[x]$ be a proposition. $\mathcal{A} \vdash_{\text{ind}} \phi[x]$ if, for all ground terms g,*

$$\mathcal{A} \cup \{\phi[g'] \mid \phi[g'] \prec \phi[g]\} \vdash \phi[g].$$

Proof. Suppose the contrary. Then there is an instance $\phi[g]$ that is not provable. Since $\prec$ is well-founded, there is a minimal $\phi[g_0] \preccurlyeq \phi[g]$ that is not provable. That would mean all $\phi[g'] \prec \phi[g_0]$ are provable. The hypothesis of the theorem then implies $\mathcal{A} \vdash \phi[g_0]$ using the axioms of $\vdash$. This gives a contradiction. □

Notation. If ϕ and ψ are propositions, we write $\{\phi\}_{\prec\psi}$ (resp. $\{\phi\}_{\preccurlyeq\psi}$) to denote the set $\{\phi\theta \mid \phi\theta \prec \psi\}$ (resp. $\{\bar{\phi}\theta \mid \phi\theta \preccurlyeq \psi\}$).

With this notation, Proposition 3 becomes

$$\mathcal{A} \vdash_{\text{ind}} \phi[x] \quad \text{if, for all ground terms } g,\ \mathcal{A} \cup \{\phi[x]\}_{\prec\phi[g]} \vdash \phi[g].$$

This proposition is at the heart of our proposal in that all of the induction methods or inductive procedures presented in this paper ultimately rely on it. To some extent, we could say that these other methods are effective variants of the above principle.

Proposition 3 is not an effective method because it involves testing all ground instances, an infinite set. To develop an effective method, we consider finite sets of nonground propositions in $\mathcal{L}$ that can *cover* all ground cases. This is formalized in the following definition.

DEFINITION (Cover sets). Let $\mathcal{A}$ be a set of axioms, and $\preccurlyeq$ a propositional order over $\mathcal{L}$. For finite sets of propositions Ψ and Φ, we say that Ψ *covers* Φ if, for every ground instance $\phi\alpha$ of a proposition $\phi \in \Phi$, $\mathcal{A} \cup (\Psi)_{\preccurlyeq\phi\alpha} \vdash \phi\alpha$. If $\Phi = \{\phi\}$ is a singleton, we call Ψ a *cover set* of ϕ.

Cover sets used in practice often have a particular $\psi\theta \in (\Psi)_{\preccurlyeq\phi\alpha}$ such that $\mathcal{A} \cup \{\psi\theta\} \vdash \phi\alpha$. (But, there is no reason to require this.)

The notion of cover sets arose in the work on completion-based induction methods and its various extensions [1, 38, 25, 23]. The present notion is a generalization of that in [31] and [10]. Automatic methods for finding cover sets for equational deduction systems may be found in the above citations. Briefly, the traditional technique with equational deduction systems is to use *narrowing* and show that the set of narrowings of a term is a cover set. Work remains to be done on finding similar methods for conditional equation systems and clausal deduction systems. (See [25].)

EXAMPLE 7. We give examples of cover sets assuming propositions are ordered as in Example 6.

1. $\{\phi\}$ is always a cover set for ϕ. If Ψ is a cover set for ϕ and Ψ' covers Ψ, then Ψ' is a cover set for ϕ. (The 'covers' relation is a preorder on sets of propositions.)
2. Let $\{t_i\}_i$ be a set of terms such that every ground term g is an instance of some t_i. Then, for any proposition ϕ, the set $\{\phi[t_i/x]\}_i$ is a cover set for ϕ. We call the set $\{t_i\}_i$ a cover set (of terms). For instance, the set $\{f(\bar{z}) \mid f \in F\}$ is a cover set.

3. Consider a deduction system with equality and let $\preccurlyeq$ be a term order that is extended to propositions as in Example 6. Suppose $\{t_i\}_i$ is a set of terms such that, for every ground term g, there is a t_i and a substitution σ such that $t_i\sigma \preccurlyeq g$ and $\mathcal{A} \vdash t_i\sigma = g$. Then, for any proposition ϕ, the set $\{\phi[t_i/x]\}_i$ is a cover set for ϕ. Again, in this case, we call $\{t_i\}$ a cover set (of terms). For instance, if there is a subset $F_0 \subseteq F$ such that, for every ground term g, there exists $g_0 \in \mathcal{T}(F_0)$ such that $g_0 \preccurlyeq g$ and $\mathcal{A} \vdash g_0 = g$, then $\{f(\bar{z})\colon f \in F_0\}$ is a cover set.
4. Consider a clausal deduction system. Suppose $\psi_1, \ldots, \psi_n$ is a collection of clauses such that each $\psi_i \prec \phi$ and $\psi_1 \vee \cdots \vee \psi_n$ is an inductive consequence of $\mathcal{A}$, i.e., $\mathcal{A} \vdash \psi_1\alpha \vee \cdots \vee \psi_n\alpha$ for all substitutions α. Then $\{\psi_1 \vee \phi, \ldots, \psi_n \vee \phi\}$ is a cover set for ϕ. (Note that $\psi_i \vee \phi \preccurlyeq_{\max} \phi$.)

PROPOSITION 4 (Noetherian induction with cover sets). *Let $\preccurlyeq$ be a propositional order. $\mathcal{A} \vdash_{\text{ind}} \phi$ if*

1. *$\{\psi_i\}_i$ is a cover set for ϕ, and*
2. *for each proposition ψ_i, we have $\mathcal{A} \cup \{\phi\}_{\prec\psi_i} \vdash^0 \psi_i$.*

(The use of $\vdash^0$ reflects that all the free variables of ψ_i are treated as eigenvariables.)

Proof. Suppose the contrary. Then there is a ground instance $\phi\alpha$ that is not provable. Since $\preccurlyeq$ is well founded, there is a minimal $\phi\alpha_0 \preccurlyeq \phi\alpha$ that is not provable. Thus, there is an instance $\psi_j\beta \preccurlyeq \phi\alpha_0$ of a proposition ψ_j of the cover set that is not provable. But hypothesis 2 implies, by axiom (v), that $\mathcal{A} \cup \{\phi\}_{\prec\psi_j\beta} \vdash \psi_j\beta$. (Stability of $\prec$ ensures that, whenever $\phi\theta \prec \psi_j$, we have $\phi\theta\beta \prec \psi_j\beta$.) So, there must be an instance $\phi\alpha' \prec \psi_j\beta$ that is not provable, contradicting the fact that $\phi\alpha_0$ is a minimal unprovable instance. □

EXAMPLE 8. Consider the following equations defining natural number addition:

$$N_1\colon \quad x + 0 = x,$$
$$N_2\colon \quad x + S(y) = S(x + y).$$

To prove the inductive theorem

$$P\colon \quad (x + y) + z = x + (y + z)$$

we use the multiset path order $\succ$ generated by the precedence $+ > S > 0$ and compare equations by $\succ_{\max}$. Every ground term is equal to a smaller (or equal) term over 0 and S. So, we can instantiate z to 0 and $S(z')$ to obtain a cover set.

1. $(x + y) + 0 \leftrightarrow_{N_1} x + y \leftrightarrow_{N_1} x + (y + 0)$,
2. $(z + y) + S(z') \leftrightarrow_{N_2} S((x + y) + z') \leftrightarrow_P S(x + (y + z')) \leftrightarrow_{N_2} x + S(y + z') \leftrightarrow_{N_2} x + (y + S(z'))$.

To show that the use of the inductive hypothesis is valid, verify that

$$\{(x+y)+z', x+(y+z')\} \prec_{\max} \{(x+y)+S(z'), x+(y+S(z'))\}.$$

A proof of this kind can be carried out in a number of proof methods and is hardly representative of the power of induction using propositional orders. The real payoff of propositional orders is obtained when we consider mutual induction proofs in the next section. A less trivial example is the following:

EXAMPLE 9. Consider a clausal deduction system with equality containing a predicate $\geqslant$ denoting a total order, a predicate $\in$ denoting membership in a list, and a function max which returns the maximum of a list. (We also assume the presence of the list constructors Nil and '$\cdot$'.) Assume the following axioms:

$$\begin{aligned}
&\mathsf{max}(a \cdot \mathsf{Nil}) = a,\\
&\mathsf{max}(a \cdot l) = \mathsf{max}(l) \Leftarrow l \neq \mathsf{Nil} \wedge \mathsf{max}(l) \geq a,\\
&\mathsf{max}(a \cdot l) = a \Leftarrow l \neq \mathsf{Nil} \wedge a \geq \mathsf{max}(l),\\
&\mathsf{max}(l) \in l \Leftarrow l \neq \mathsf{Nil}.
\end{aligned}$$

To verify the correctness of the definition of max, we wish to prove

$$P[l]: \quad \mathsf{max}(l) \geq x \Leftarrow l \neq \mathsf{Nil} \wedge x \in l.$$

To order terms, we use the multiset path order with the precedence order max $>$ '$\in$' $>$ '$\geq$' $>$ '$\cdot$' $>$ Nil and extend it to clauses as described in Example 6. The following four cases form a cover set of $P[l]$. (First, $\{P[\mathsf{Nil}], P[a \cdot l']\}$ covers $P[l]$ by Example 6, item 3. $\{P[a \cdot l'] \Leftarrow l' = \mathsf{Nil},\ P[a \cdot l'] \Leftarrow l' \neq \mathsf{Nil}\}$ covers $P[a \cdot l']$ by Example 6, item 4. Use this item again with $\mathsf{max}(l') \geqslant a \vee a \geqslant \mathsf{max}(l')$ to obtain the last two cases.)

1. $P[\mathsf{Nil}]$ reduces to true because $\mathsf{Nil} \neq \mathsf{Nil}$ is false.
2. $P[a \cdot l] \Leftarrow l' = \mathsf{Nil}$ reduces to true because $a \geq a$ is true.
3. $P[a \cdot l'] \Leftarrow l' \neq \mathsf{Nil} \wedge \mathsf{max}(l') \geq a$ reduces to

$$Q_1: \quad \mathsf{max}(l') \geq x \Leftarrow l' \neq \mathsf{Nil} \wedge \mathsf{max}(l') \geq a \wedge (x = a \vee x \in l').$$

The case $x = a$ is trivial, while the case $x \in l'$ follows from the inductive hypothesis $P[l']$ (which is smaller than $P[a \cdot l']$).

4. $P[a \cdot l'] \Leftarrow l' \neq \mathsf{Nil} \wedge a \geq \mathsf{max}(l')$ reduces to

$$Q_2: \quad a \geq x \Leftarrow l' \neq \mathsf{Nil} \wedge a \geq \mathsf{max}(l') \wedge (x = a \vee x \in l').$$

The case $x = a$ reduces to true because $a \geq a$ is true. The case $x \in l'$ follows from the following resolvent of $P[l']$:

$$\frac{z \geq x \Leftarrow z \geq u \wedge u \geq x \quad \mathsf{max}(l') \geq x \Leftarrow l' \neq \mathsf{Nil} \wedge x \in l'}{z \geq x \Leftarrow z \geq \mathsf{max}(l') \wedge l' \neq \mathsf{Nil} \wedge x \in l'} \text{ Resolution.}$$

While this second example appears to be quite simple, implicit induction methods have trouble duplicating it because the transitivity axiom is not used *reductively*. As we show later, implicit induction procedures insist on using axioms reductively, so they fail to prove this property without additional lemmas.

4. Mutual Induction

The major benefit of using propositional orders rather than semantic orders for inductive reasoning is that it allows proof by *mutual induction*, that is, proofs of multiple propositions that use each other in their proofs in a mutually recursive fashion. For example, suppose we have two propositions $\phi[x]$ and $\psi[y]$, a propositional order $\preccurlyeq$, and we show that for all ground terms g and g',

$$\mathcal{A} \cup \{\phi[x]\}_{\prec\phi[g]} \cup \{\psi[y]\}_{\prec\phi[g]} \vdash \phi[g], \tag{1}$$

$$\mathcal{A} \cup \{\phi[x]\}_{\prec\psi[g']} \cup \{\psi[y]\}_{\prec\psi[g']} \vdash \psi[g']. \tag{2}$$

Then, both $\phi[x]$ and $\psi[y]$ are inductive theorems of $\mathcal{A}$. For, if a proof of $\phi[g]$ by (1) uses an instance $\psi[t]$ of $\psi[y]$, then $\psi[t] \prec \phi[g]$. And if the proof of $\psi[t]$ by (2) uses an instance $\phi[t']$ of $\phi[x]$, then $\phi[t'] \prec \psi[t]$. Thus, $\phi[t'] \prec \phi[g]$. Hence, such mutually inductive proofs, where the proof of $\phi[x]$ can use $\psi[y]$ and the proof of $\psi[y]$ can use $\phi[x]$, are valid.

In contrast, a classical induction proof to be valid would have to be of the form

$$\mathcal{A} \cup \{(z < x \Rightarrow \phi[z]), (z < x \Rightarrow \psi[u])\} \vdash^0 \phi[x] \wedge \psi[x].$$

The comparative strength of induction using propositional orders to realize such induction proof come from its flexibility. For example, x in $\phi[x]$ and y in $\psi[y]$ do not have to be of the same type. Similarly, when cover sets are used, the cover sets selected for $\phi[x]$ and $\psi[y]$ do not have to be the same. These effects would be harder to achieve in the setting of classical induction.

The following procedure formalizes proof by mutual induction:

PROCEDURE 1 (Mutual induction procedure). Let $\preccurlyeq$ be a propositional order and $\mathcal{A}$ a set of axioms. The *mutual induction procedure* is defined by the following transition rules applied to pairs of the form (P, H) where P is the set of inductive propositions and H the set of inductive hypotheses:

$$\begin{array}{ll} \textit{Expand} & (P \cup \{\phi\}, H) \Rightarrow (P \cup \Psi, H \cup \{\phi\}) \quad \text{if } \Psi \text{ is a cover set of } \phi. \\ \textit{Delete} & (P \cup \{\phi\}, H) \Rightarrow (P, H) \quad \text{if } \mathcal{A} \cup (P)_{\preccurlyeq\phi} \cup (H)_{\prec\phi} \vdash^0 \phi. \\ \textit{Lemma} & (P, H) \Rightarrow (P \cup P', H). \end{array}$$

A successful derivation is a sequence of transitions, $(P_0, \emptyset) \Rightarrow \cdots \Rightarrow (P_n, H_n)$, such that $P_n = \emptyset$.

In practice, we also use a derived transition rule that combines *Delete* and *Lemma*:

$$\begin{array}{ll} \mathit{Simplify} & (P \cup \{\phi\}, H) \Rightarrow (P \cup P', H) \\ & \text{if } \mathcal{A} \cup (P \cup P')_{\preccurlyeq \phi} \cup (H)_{\prec \phi} \vdash^0 \phi. \end{array}$$

This rule means that if we can reduce a proposition ϕ to a set of propositions P' then these latter propositions can be adopted as lemmas and we can attempt a mutual induction proof of the lemmas and together with the initial propositions. In contrast, a classical induction method would need either to prove these lemmas independently or to redo the whole proof after combining the lemmas and the initial propositions. The strength of induction using propositional orders comes here from the flexibility that such *on-the-fly* construction of mutual induction proofs allows.

EXAMPLE 10. Consider the Equations N_1, N_2 given in Example 8 defining natural number addition. We wish to prove the inductive theorem

$$P_0\colon \quad x + y = y + x.$$

To prove this theorem, we instantiate y using the cover set $\{0, S(y')\}$ and reduce the resulting equations to normal form:

$$\begin{array}{ll} P_1\colon & (x + 0 = 0 + x) \leftrightarrow_{N_1} x = 0 + x, \\ P_2\colon & (x + S(y') = S(y') + x) \leftrightarrow_{N_2} S(x + y') = S(y') + x. \end{array}$$

The reduced forms can be proved as lemmas. For P_1 the proof is straightforward and resembles Example 8. For P_2, we instantiate x using the cover set $\{0, S(x')\}$ and obtain

$$\begin{array}{l} S(0 + y') \leftrightarrow_{P_1} S(y') \leftrightarrow_{N_1} S(y') + 0, \\ S(S(x') + y') \leftrightarrow_{P_2} S(S(y' + x')) \leftrightarrow_{P_0} S(S(x' + y')) \\ \quad \leftrightarrow_{P_2} S(S(y') + x') \leftrightarrow_{N_2} S(y') + S(x'). \end{array}$$

Note the mutual use of the lemmas and the main theorem here. The proof of P_0 requires Lemma P_2; in turn, the proof of P_2 needs P_0 to show $y' + x' = x' + y'$. The use of the inductive hypothesis corresponding to P_0 is valid because

$$\{y' + x', x' + y'\} \prec_{\max} \{S(S(x') + y'), S(y') + S(x')\}.$$

Notice, in this example, the phenomenon mentioned in the Introduction. If we prove this theorem by hand, we might formulate a lemma of the form $S(y') + x = S(y' + x)$ which has an independent inductive proof. On the other hand, the automatically generated Lemma P_2 is dependent on the theorem for its proof, requiring mutual induction.

4.1. Soundness of Mutual Induction

THEOREM 5 (Soundness of mutual induction). *If* $(P_0, \emptyset) \Rightarrow \cdots \Rightarrow (\emptyset, H_n)$ *is a derivation of the mutual induction procedure, then* $\mathcal{A} \vdash_{\text{ind}} P_0$.

To prove this theorem, we rely on the following lemma:

LEMMA 6. *Let* $(P_0, \emptyset) \Rightarrow \cdots \Rightarrow (P_i, H_i) \Rightarrow \cdots \Rightarrow (\emptyset, H_n)$ *be a derivation of the mutual induction procedure. For all* $i = 0, \ldots, n$, *the following invariants hold:*

$$\forall \phi \in \bigcup_{j=0,\ldots,i} P_j,\ \forall \textit{ ground } \alpha, \quad \mathcal{A} \cup (P_i)_{\preccurlyeq \phi\alpha} \vdash \phi\alpha, \tag{3}$$

$$\forall \phi \in H_i,\ \forall \textit{ ground } \alpha, \quad \mathcal{A} \cup (P_i)_{\preccurlyeq \phi\alpha} \vdash \phi\alpha. \tag{4}$$

Invariant (3) states that each P_i covers the earlier P_j's, while (4) states that P_i covers H_i.

Proof. For $i = 0$, the invariants hold because P_0 covers itself and H_0 is empty.

For $i > 0$, $(P_{i-1}, H_{i-1}) \Rightarrow (P_i, H_i)$ is an instance of one of the transition rules. The rules *Expand* and *Lemma* are straightforward. For the *Expand* rule, (3) follows from Ψ being a cover set of ϕ, and (4) follows from applying the inductive hypotheses to show that $P \cup \Psi$ covers H.

To show that *Delete* maintains (3), we need to show $\mathcal{A} \cup (P)_{\preccurlyeq \phi\alpha} \vdash \phi\alpha$. By the hypothesis of the rule, $\mathcal{A} \cup (P)_{\preccurlyeq \phi\alpha} \cup (H)_{\prec \phi\alpha} \vdash \phi\alpha$ (since the order is stable). By inductive hypothesis (4), we have $\mathcal{A} \cup (P \cup \{\phi\})_{\preccurlyeq \phi'\alpha'} \vdash \phi'\alpha'$ for every ground instance $\phi'\alpha'$ of a proposition in H. Thus $P \cup \{\phi\}$ covers H. So, by the hypothesis of the rule, $\mathcal{A} \cup (P)_{\preccurlyeq \phi\alpha} \cup (P \cup \{\phi\})_{\prec \phi\alpha} \vdash \phi\alpha$. This is equivalently written as $\mathcal{A} \cup (P)_{\preccurlyeq \phi\alpha} \cup \{\phi\}_{\prec \phi\alpha} \vdash \phi\alpha$. By Proposition 3, $\mathcal{A} \cup (P)_{\preccurlyeq \phi\alpha} \vdash \phi\alpha$. Thus, P cover $P \cup \{\phi\}$. The invariant (4) now follows, since P covers $P \cup \{\phi\}$, which, in turn, covers H. □

Invariant (3) applied to $i = n$ gives that P_n $(= \emptyset)$ covers P_0. Thus, $\mathcal{A} \vdash_{\text{ind}} P_0$, proving Theorem 5. □

An impressive example of the power of mutual induction is the proof of the Gilbreath card trick [4]. Proofs produced by the Coq, Nqthm, and RRL systems, which use classical induction, require approximately fifteen to twenty lemmas. In contrast, systems based on mutual induction (such as SPIKE and Focus) need only two to ten lemmas. More importantly, finding some of the lemmas used in the classical proof requires significant insight. In contrast, the lemmas used by the proofs by mutual induction are straightforward and their discovery would be easy to automate.

We give another example of the power of mutual induction in the Appendix. This example involves showing the correctness of a simple unification algorithm.

A similar theorem was proven by Rouyer and Lescanne using the Coq proof assistant with conventional induction [35]. Their proof requires 45 definitions and 48 lemmas and theorems. In contrast, our proof requires 14 definitions and 17 lemmas and theorems. However, we did not include a proof of the well-foundedness of our term order while Rouyer and Lescanne's proof includes such a proof.

The difference between the two proofs is interesting. To prove the correctness of the unification algorithm, one must prove the correctness of two separate unification functions: one for terms and one for lists of terms. The difficulty is that both the properties need each other in their proofs although their parameters are of different types. This is not a problem in our approach because of the use of term orders. In contrast, it is unclear what to do in the conventional setting. Rouyer and Lescanne introduced the type of *quasi-terms* to combine both terms and lists of terms into one type. Then, they solved the unification problem for quasi-terms and showed that this implied that the unification problem for terms was solved as well.

5. Backward Inferences

The rule *Delete* involves producing a proof

$$\mathcal{A} \cup (P \cup H)_{\prec\phi} \vdash^0 \phi$$

for a proposition ϕ. If the deduction system works *backwards*, i.e., reduces the goal proposition ϕ to a series of goals $\psi_1, \ldots, \psi_n$ until a trivial goal ψ_n is obtained, one can test for the condition '$\prec \phi$' for each premise used in deriving a goal ψ_i from ψ_{i+1}. If, on the other hand, the deduction system also involves *forward* inferences, whereby new consequences are derived from existing premises in $\mathcal{A} \cup P \cup H$, we need a way to transform the condition '$\prec \phi$' to appropriate conditions on the consequences. We study these issues in the following two sections.

DEFINITION. A *subgoal proposition* is a pair of propositions (ρ, ψ), written $\rho \leftleftarrows \psi$, such that $\psi \preccurlyeq \rho$. The propositions ρ and ψ are called the *goal* and *subgoal* of $\rho \leftleftarrows \psi$, respectively.

The idea is that when we reduce a goal proposition ρ to a subgoal ψ, we still keep track of the goal ρ and use it in induction. The condition $\psi \preccurlyeq \rho$ is not strictly necessary, but it is convenient to have.

If $\rho \leftleftarrows \psi$ is a subgoal proposition and ϕ a proposition, $\{\rho \leftleftarrows \psi\}_{\preccurlyeq\phi\leftleftarrows}$ (resp. $\{\rho \leftleftarrows \psi\}_{\prec\phi\leftleftarrows}$) denotes the set of propositions $\{\psi\theta \mid \rho\theta \preccurlyeq \phi\}$ (resp. $\{\psi\theta \mid \rho\theta \prec \phi\}$). Note that this parallels the notation used for ordinary propositions. Moreover, $\{\psi \leftleftarrows \psi\}_{\preccurlyeq\phi\leftleftarrows}$ is nothing but $\{\psi\}_{\preccurlyeq\phi}$.

Proposition 3 is adapted to subgoal propositions as follows:

PROPOSITION 7 (Noetherian induction for subgoal propositions). *If, for every grounding substitution* α, $\mathcal{A} \cup \{\rho \Leftarrow \phi\}_{\prec \rho\alpha \Leftarrow} \vdash \phi\alpha$ *then* $\mathcal{A} \vdash_{\text{ind}} \phi$.

Proof. Suppose the contrary. Then there is an instance $\phi\alpha$ that is not provable. Since $\prec$ is well founded, there is a minimal $\rho\alpha_0 \preccurlyeq \rho\alpha$ such that $\phi\alpha_0$ is not provable. But, since all instances $\{\phi\alpha' \mid \rho\alpha' \prec \rho\alpha_0\}$ are provable, $\phi\alpha_0$ must be provable. □

To adapt the mutual induction procedure to subgoal propositions, we must decide how to build subgoals propositions from ordinary propositions. Propositions arising in cover sets can be treated as both goals and subgoals, i.e., as subgoals propositions of the form $\psi \Leftarrow \psi$. (We also write $\Psi \Leftarrow \Psi$ to mean $\{\psi \Leftarrow \psi\colon\ \psi \in \Psi\}$.) Similarly, the initial conjectures and the lemmas added in the course of the procedure can also be treated as both goals and subgoals, in which case they form subgoal propositions of the form $\psi \Leftarrow \psi$. However, for these conjectures a greater flexibility is possible. With a conjecture ϕ, one could, if desired, associate any proposition ρ, such that $\rho \succeq \phi$. We denote this by using a function Δ such that $\Delta(\phi) = \rho \Leftarrow \phi$, ρ is some proposition $\succeq \phi$. Δ is extended to sets of propositions by $\Delta\Phi = \{\Delta\phi \mid \phi \in \Phi\}$.

PROCEDURE 2 (Explicit induction procedure for backward inferences). Let $\preccurlyeq$ be a propositional order and $\mathcal{A}$ a set of axioms. The *explicit induction procedure* is defined by the following transition rules applied to pairs of the form (P, H) where P is a set of subgoal propositions (of conjectures to be proved) and H a set of propositions (inductive hypotheses):

$$\begin{array}{lll} \textit{Expand} & (P \cup \{\rho \Leftarrow \phi\}, H) \Rightarrow & (P \cup (\Psi \Leftarrow \Psi), H \cup \{\phi\}) \\ & & \text{if } \Psi \text{ is a cover set of } \phi. \\ \textit{Simplify} & (P \cup \{\rho \Leftarrow \phi\}, H) \Rightarrow & (P \cup P', H) \\ & & \text{if } \mathcal{A} \cup (P \cup P')_{\preccurlyeq \rho \Leftarrow} \cup (H)_{\prec \rho} \vdash^0 \phi. \\ \textit{Lemma} & (P, H) \Rightarrow (P \cup \Delta(\Phi), H). & \end{array}$$

A successful derivation is a sequence of transitions, $(P_0, \emptyset) \Rightarrow \cdots \Rightarrow (P_n, H_n)$, such that $P_0 = \Delta(C)$ for some set of propositions C and $P_n = \emptyset$.

THEOREM 8 (Soundness of explicit induction). *If* $(\Delta(C), \emptyset) \Rightarrow \cdots \Rightarrow (\emptyset, H_n)$ *is a derivation of the explicit induction procedure, then* $\mathcal{A} \vdash_{\text{ind}} C$.

To prove this theorem, we rely on the following lemma:

LEMMA 9. *Let* $(P_0, \emptyset) \Rightarrow \cdots \Rightarrow (P_i, H_i) \Rightarrow \cdots \Rightarrow (\emptyset, H_n)$ *be a derivation of the explicit induction procedure. For all* $i = 0, \ldots, n$, *the following invariants hold:*

$$\forall (\rho \Leftarrow \phi) \in \bigcup_{j=0,\ldots,i} P_j,\ \forall \textit{ ground } \alpha, \quad \mathcal{A} \cup (P_i)_{\preccurlyeq \rho\alpha \Leftarrow} \vdash \phi\alpha, \tag{5}$$

$$\forall \phi \in H_i,\ \forall \textit{ ground } \alpha, \quad \mathcal{A} \cup (P_i)_{\preccurlyeq \phi\alpha \Leftarrow} \vdash \phi\alpha. \tag{6}$$

For sets of subgoal propositions P and Q, say that P *covers* Q if $\mathcal{A} \cup (P)_{\preccurlyeq \rho\alpha \Leftarrow} \vdash \psi\alpha$ for all ground instances $\rho\alpha \Leftarrow \psi\alpha$ of subgoal propositions in Q. Then, the covers relation is a preorder. Further, $\Psi \Leftarrow \Psi$ covers $\Phi \Leftarrow \Phi$ whenever Ψ covers Φ as a set of propositions. Now, invariant (5) states that each P_i covers the earlier P_j's, while (6) states that P_i covers $H_i \Leftarrow H_i$.

Proof. For $i = 0$, the invariants hold because P_0 covers itself and H_0 is empty.

For $i > 0$, $(P_{i-1}, H_{i-1}) \Rightarrow (P_i, H_i)$ is an instance of one of the transition rules. For the *Expand* rule, (5) follows from the fact Ψ covers ϕ, i.e., $\mathcal{A} \cup (\Psi)_{\preccurlyeq \phi\alpha} \vdash \phi\alpha$, and, since $\phi\alpha \preccurlyeq \rho\alpha$, $\mathcal{A} \cup (\Psi)_{\preccurlyeq \rho\alpha} \vdash \phi\alpha$. The latter fact means that $\Psi \Leftarrow \Psi$ covers $\rho \Leftarrow \phi$. The invariant (6) follows from applying the inductive hypotheses to show that $P \cup (\Psi \Leftarrow \Psi)$ covers $H \Leftarrow H$.

To show that *Simplify* maintains (5), we need to show that $P \cup P'$ covers $\rho \Leftarrow \phi$, i.e., $\mathcal{A} \cup (P \cup P')_{\preccurlyeq \rho\alpha \Leftarrow} \vdash \phi\alpha$. By the hypothesis of the rule, $\mathcal{A} \cup (P \cup P')_{\preccurlyeq \rho\alpha \Leftarrow} \cup (H)_{\prec \rho\alpha} \vdash \phi\alpha$. By inductive hypothesis (6), $P \cup \{\rho \Leftarrow \phi\}$ covers $H \Leftarrow H$. So, $\mathcal{A} \cup (P \cup P')_{\preccurlyeq \rho\alpha \Leftarrow} \cup (P \cup \{\rho \Leftarrow \phi\})_{\prec \rho\alpha \Leftarrow} \vdash \phi\alpha$. This is equivalently written as $\mathcal{A} \cup (P \cup P')_{\preccurlyeq \rho\alpha \Leftarrow} \cup \{\rho \Leftarrow \phi\}_{\prec \rho\alpha \Leftarrow} \vdash \phi\alpha$. By Proposition 7, $\mathcal{A} \cup (P \cup P')_{\preccurlyeq \rho\alpha \Leftarrow} \vdash \phi\alpha$. Thus, $P \cup P'$ covers $\rho \Leftarrow \phi$. The invariant (6) now follows, since $P \cup P'$ covers $P \cup \{\rho \Leftarrow \phi\}$, which, in turn, covers $H \Leftarrow H$. □

EXAMPLE 11. Consider the following equational theory from [31].

$$
\begin{aligned}
F1:&\quad f(0) = 0,\\
F2:&\quad f(S0) = S0,\\
F3:&\quad f(SSn) = f(Sn) + f(n),\\
G1:&\quad g(0) = \langle S0, 0\rangle,\\
G2:&\quad g(Sn) = np(g(n)),\\
G3:&\quad np(\langle x, y\rangle) = \langle x + y, x\rangle.
\end{aligned}
$$

We would like to prove the theorem

$$P[x]:\quad g(x) = \langle f(Sx), f(x)\rangle$$

using the recursive path order generated by the precedence

$$f > g > np > \langle -, - \rangle > + > S > 0.$$

Use the cover set $\{P[0], P[Sn]\}$. The case $P[0]$ is straightforward, while $P[Sn]$ reduces to the subgoal proposition

$$P[Sn] \Leftarrow np(g(n)) = \langle f(Sn) + f(n), f(Sn)\rangle.$$

We may now use the inductive hypothesis $P[n]$, which is smaller than $P[Sn]$, to obtain

$$P[Sn] \Leftarrow np(\langle f(Sn), f(n)\rangle) = \langle f(Sn) + f(n), f(Sn)\rangle$$

and complete the proof using $G3$. The interesting point is that the indicated step is not a reduction: $np(g(n)) \not\prec np(\langle f(Sn), f(n)\rangle)$. Such a step would not be permissible in an implicit induction procedure. In fact, the typical implicit induction procedure diverges for this example [31].

6. Forward Inferences

In this section, we assume that the deduction relation $\vdash$ is composed of relations $\vdash_F$ and $\vdash_B$ representing forward and backward deductions, respectively, and show how to adapt the mutual induction procedure in that context.

A forward inference of the form

$$\frac{\psi_1 \cdots \psi_n}{\psi}$$

(where all propositions have been renamed apart) typically uses unification to derived the conclusion ψ. In that case, there is a substitution θ such that

$$\frac{\psi_1\theta\alpha \cdots \psi_n\theta\alpha}{\psi\alpha}$$

is also a forward inference for all grounding substitutions α. This motivates the following definition.

DEFINITION. Consider relations $F \subseteq \mathrm{Mul}(\mathcal{L}) \times \mathcal{L}$ between multisets of propositions and propositions, where all propositions are renamed apart. Call a relation F a *forward inference* relation if, for each $(\Psi, \psi) \in F$ there is a substitution θ (called the underlying substitution of the inference) such that, for all ground substitutions α, $(\Psi\theta\alpha, \psi\alpha) \in F$. The corresponding *forward deduction* relation $\vdash_F$ is the least deduction relation such that $\mathcal{A} \vdash_F \psi$ whenever $(\Psi, \psi) \in F$ with every proposition of Ψ in $\mathcal{A}$.

EXAMPLE 12. For equational deduction, a standard forward deduction mechanism is superposition (generation of critical pairs). Assuming $\prec$ is a term order, superposition is expressed by the following rule:

$$\frac{r = s \quad t = u}{t[s\theta/\alpha] = u\theta} \quad \text{if} \quad \theta = \mathrm{mgu}(r, t/\alpha),\ t/\alpha \text{ non} - \text{variable},$$
$$r\theta \not\prec s\theta \text{ and } t\theta \not\prec u\theta.$$

If the order $\prec$ is total on ground terms, superposition and (ordered) rewriting (a backward deduction mechanism) form a complete semi-decision procedure for proving equations [2, 19].

EXAMPLE 13. For clausal reasoning, one may use the following *critical resolution* as a forward deduction mechanism. Assuming $\prec$ is a term order on literals,

$$\frac{a \vee C \quad \neg a' \vee C'}{C\theta \vee C'\theta} \quad \text{if } \theta = \text{mgu}(a, a'), \{a\} \not\prec_{\text{mul}} C \text{ and } \{\neg a'\} \not\prec_{\text{mul}} C'.$$

If the order $\prec$ is total on ground literals, critical resolution and reductive deduction form a complete semi-decision procedure for proving clauses [7, 9].

Since the propositions in $P \cup H$ can be used only in *Delete* in a restricted context, we need to track the dependencies on such propositions through forward inferences. We use the following representation to keep track of these dependencies.

DEFINITION. A *supported proposition* is a pair written $S \rightrightarrows \psi$ where ψ is a proposition and $S = \{\zeta_1, \ldots, \zeta_n\}$ is a finite set of propositions (called the *support* of ψ). We often write $\{\zeta_1, \ldots, \zeta_n\} \rightrightarrows \psi$ as $\zeta_1, \ldots, \zeta_n \rightrightarrows \psi$ and $\emptyset \rightrightarrows \psi$ as simply ψ.

The forward consequences of inductive hypotheses are tagged with the instances of the inductive hypotheses that they depend on. Specifically, for every supported proposition $S \rightrightarrows \psi$, we ensure that

$$\forall \text{ ground } \alpha, \quad \mathcal{A} \cup S\alpha \vdash \psi\alpha.$$

A fresh inductive hypothesis has the form $\psi \rightrightarrows \psi$, while an axiom ψ of $\mathcal{A}$ is trivially regarded as a supported proposition $\emptyset \rightrightarrows \psi$.

If P is a set of supported propositions, we use the notation

$$P_{\prec\phi\rightrightarrows} = \{\psi\theta \mid (S \rightrightarrows \psi) \in P, \zeta\theta \prec \phi \text{ for all } \zeta \in S\}$$

to denote all instances of P derivable from the instances of hypotheses admissible for ϕ. Note that $\{\psi \rightrightarrows \psi\}_{\prec\phi\rightrightarrows} = \{\psi\}_{\prec\phi}$.

DEFINITION. If F is a forward inference relation, the corresponding *supported* inference relation $F_{\rightrightarrows}$ consists of inferences

$$(\{S_1 \rightrightarrows \psi_1, \ldots, S_n \rightrightarrows \psi_n\}, S_1\theta \cup \cdots \cup S_n\theta \rightrightarrows \psi)$$

whenever $(\{\psi_1, \ldots, \psi_n\}, \psi)$ is an inference in F with the underlying substitution θ. (That is, the conclusion inherits the support of all the premises after instantiating them appropriately.)

Note that the definition gives an effective mechanism for transforming the inference rules of $\vdash_F$ to those of $\vdash_{F_{\rightrightarrows}}$.

For example, the supported version of the resolution inference required in Example 9 is

$$\frac{z \geq x \Leftarrow z \geq u \wedge u \geq x \quad P[l'] \rightrightarrows P[l']}{P[l'] \rightrightarrows (z \geq x \Leftarrow z \geq \mathsf{max}(l') \wedge l' \neq \mathsf{Nil} \wedge x \in l')} \text{ Resolution.}$$

The support $P[l']$ of the conclusion indicates that it may be used in the proof of $P[a \cdot l']$.

Similarly, for the problem of Example 11, there exists a supported superposition inference:

$$\frac{np(\langle x, y\rangle) = \langle x + y, x\rangle \quad P[n] \rightrightarrows P[n]}{P[n] \rightrightarrows np(g(n)) = \langle f(Sn) + f(n), f(Sn)\rangle}.$$

The conclusion can be used to directly prove the cover set instance $P[Sn]$ instead of using a nonreductive backward inference.

PROCEDURE 3 (Explicit induction procedure). Let $\mathcal{A}$ be a theory, $\vdash_B$ a (backward) deduction relation, F a forward inference relation and $F_{\rightrightarrows}$ the supported version of F. The *explicit induction procedure* is defined by the following transition rules applied to pairs of the form (P, H) where P is the set of conjectures and H the set of inductive hypotheses:

$$\begin{array}{ll} \textit{Expand} & (P \cup \{\rho \leftleftarrows \phi\}, H) \Rightarrow (P \cup (\Psi \leftleftarrows \Psi), H \cup \{\phi \rightrightarrows \phi\}) \\ & \qquad \text{if } \Psi \text{ is a cover set for } \phi. \\ \textit{Simplify} & (P \cup \{\rho \leftleftarrows \phi\}, H) \Rightarrow (P \cup P', H) \\ & \qquad \text{if } \mathcal{A} \cup (P \cup P')_{\preccurlyeq\rho\leftleftarrows} \cup (H)_{\prec\rho\rightrightarrows} \vdash^0_B \phi. \\ \textit{Forward} & (P, H) \Rightarrow (P, H \cup \{S \rightrightarrows \phi\}) \\ & \qquad \text{if } \mathcal{A} \cup H \vdash_{F_{\rightrightarrows}} (S \rightrightarrows \phi). \\ \textit{Lemma} & (P, H) \Rightarrow (P \cup \Delta(\Phi), H). \end{array}$$

THEOREM 10 (Soundness). *If* $(\Delta(C), \emptyset) \Rightarrow \cdots \Rightarrow (\emptyset, H_n)$ *is a derivation of the explicit induction procedure, then* $\mathcal{A} \vdash_{\mathrm{ind}} C$.

The proof of this is similar to that of Theorem 5 based on the following invariants:

$$\forall(\rho \leftleftarrows \phi) \in \bigcup_{j=0,\ldots,i} P_j,\ \forall \text{ ground } \alpha, \quad \mathcal{A} \cup (P_i)_{\preccurlyeq\phi\alpha\leftleftarrows} \vdash \phi\alpha, \tag{7}$$

$$\forall(S \rightrightarrows \phi) \in H_i, \forall\psi \in S,\ \forall \text{ ground } \alpha, \quad \mathcal{A} \cup (P_i)_{\preccurlyeq\psi\alpha\leftleftarrows} \vdash \psi\alpha, \tag{8}$$

$$\forall(S \rightrightarrows \phi) \in H_i,\ \forall \text{ ground } \alpha, \quad \mathcal{A} \cup S\alpha \vdash \phi\alpha. \tag{9}$$

Here, $\vdash$ is the least deduction relation containing $\vdash_B \cup \vdash_F$. Note that (8) is an adaptation of the invariant (4) of the mutual induction procedure. Invariant (9) is new. It represents the semantics of support.

7. Relation to Implicit Induction

In the original completion-based induction method and all subsequent *implicit induction* methods, the backward inferences are carried out using a *reductive* deduction mechanism. Such reductive deductions always reduce goal propositions to smaller propositions by the given propositional order. In this context, explicit induction tests and some of the bookkeeping involved in the explicit induction procedure is redundant. In this section, we study the use of reductive deduction methods and relate our procedure to the traditional implicit induction procedures.

DEFINITION (Reductive deduction relation). A deduction relation $\vdash_R$ is *reductive* (with respect to a propositional order $\preccurlyeq$) if

$$\mathcal{A} \vdash_R \phi \Leftrightarrow \mathcal{A}_{\preccurlyeq\phi} \vdash_R^0 \phi.$$

So, a reductive inference uses instances of axioms smaller than the goal proposition ϕ. Moreover, if it simplifies ϕ to a collection of subgoals Ψ, each proposition in Ψ would be smaller than ϕ.

EXAMPLE 14. Term rewriting [24] is a reductive deduction relation. A terminating rewrite system $\mathcal{A}$ is a set of rewrite rules with a well-founded reduction order $\prec$ on $\mathcal{T}$ such that, for each $(a \to b) \in \mathcal{A}$, $a \succ b$. Assume, without loss of generality, that $\prec$ includes the (strict) subterm order (as any reduction order is compatible with it). The rewrite relation $\to_\mathcal{A}$ is defined by $t[a\sigma/\alpha] \to_\mathcal{A} [b\sigma/\alpha]$ where α is a position in $t, a \to b$ a rewrite rule in $\mathcal{A}$ and σ a substitution. Consider the term t in the equation $t = u$. It is easy to see that whenever $t \to_\mathcal{A} t'$ by an instance $a\sigma \to b\sigma$ of a rewrite rule, $\{t, u\} \succcurlyeq_{\max} \{a\sigma, b\sigma\}$. More generally, given a rewrite proof $t \to_\mathcal{A}^* v \leftarrow_\mathcal{A}^* u$ of $t = u$, each instance of the rewrite rule used in the proof is smaller than $\{t, u\}$ by $\preccurlyeq_{\max}$. Thus, the deduction system of term rewriting is reductive.

Similarly, conditional rewriting [32] is a reductive mechanism by the order on conditional equations defined in Example 6. Likewise, reductive deduction for clauses [9] is a reductive mechanism by the order on clauses.

If one uses reductive deduction relations, then the induction method of Proposition 4 can be simplified as follows:

DEFINITION (Strict cover sets). Let $\mathcal{A}$ be a set of axioms and $\preccurlyeq$ a propositional order. We say that Ψ is a *strict cover set* for ϕ if, for all ground instances $\phi\alpha$, $\mathcal{A} \cup (\Psi)_{\prec\phi\alpha} \vdash \phi\alpha$.

PROPOSITION 11 (Implicit induction principle). *Let $\vdash_R$ be a reductive deduction relation. $\mathcal{A} \vdash_{\text{ind}} \phi$ if,*

1. *$\{\psi_i\}_i$ is a strict cover set for ϕ, and*
2. *for each proposition ψ_i, we have $\mathcal{A} \cup \{\phi\} \vdash_R \psi_i$.*

The main difference with Proposition 3 is that the strict decrease by the order $\prec$ is transferred from step 2 to step 1. Hence, in step 2, instances $\{\phi\}_{\preceq\psi_i\alpha}$ can be used in proving each $\psi_i\alpha$ and, the fact that $\vdash_R$ is *reductive* guarantees that only such instances are used. Thus, the test for appropriate instances of inductive hypothesis is 'implicit' in the deduction mechanism.

A mutual induction procedure can be devised based on this principle as follows:

PROCEDURE 4 (Implicit induction procedure). Let $\mathcal{A}$ be a set of axioms and $\vdash_R$ be a reductive deduction relation. The *implicit induction procedure* is defined by the following transition rules:

$$\begin{array}{ll} \mathit{Expand} & (P \cup \{\phi\}, H) \Rightarrow (P \cup \Psi, H \cup \{\phi\}) \\ & \qquad\qquad\qquad \text{if } \Psi \text{ is a strict cover set for } \phi. \\ \mathit{Simplify} & (P \cup \{\phi\}, H) \Rightarrow (P \cup P', H) \\ & \qquad\qquad\qquad \text{if } \mathcal{A} \cup P \cup P' \cup H \vdash_R \phi. \\ \mathit{Lemma} & (P, H) \Rightarrow (P \cup P', H). \end{array}$$

THEOREM 12 (Soundness of implicit induction). *If* $(P_0, \emptyset) \Rightarrow \cdots \Rightarrow (\emptyset, H_n)$ *is a derivation of the implicit induction procedure, then* $\mathcal{A} \vdash_{\text{ind}} C$.

The proof of this theorem is similar to the soundness proof of the mutual induction procedure. The main difference is that in the counterpart of Lemma 6, we would replace the second invariant by the following stronger invariant:

$$\forall \phi \in H_i,\ \forall \text{ ground } \alpha, \quad \mathcal{A} \cup (P_i)_{\prec\phi\alpha} \vdash \phi\alpha. \tag{10}$$

This invariant is preserved in *Expand* steps because of the use of strict cover sets. In turn, it allows in *Simplify* steps the weaker constraint $H_{\preceq\phi}$.

Instances of the implicit induction procedure include the hierarchical induction procedure defined by Reddy [31] where $\vdash_R$ is the rewriting deduction mechanism, the inductive procedures for conditional equations of Kounalis and Rusinowitch [25], Bronsard and Reddy [8], and Bouhoula and Rusinowitch [4].

Although the implicit induction procedure relies on a stronger simplification mechanism than Procedure 3 – it can use instances of propositions from H equal to ϕ – the implicit induction procedure is essentially subsumed by Procedure 3. If ϕ is in the strict cover set of some proposition ρ, then the subgoal proposition corresponding to ϕ in the explicit procedure would normally be of the form $\rho' \Leftarrow \phi$ with $\rho' \succ \phi$. Thus, using an instance of H equal to ϕ would be allowed by Procedure 3. This assumes, however, that corresponding to any strict cover set Ψ of ρ, we can find a non-strict cover set Ψ' so that Ψ' is reducible to Ψ. This is indeed the case with most methods for finding cover sets. Finally, if ϕ is part of the initial set P_0 or introduced as a lemma, then an appropriate function Δ can be defined to ensure that the goal associated with ϕ is greater than ϕ. Hence the instances used to simplify ϕ would be less than that goal and the simplification would be valid.

The limitation of the implicit induction procedure is that only *reductive* deduction relations can be used in the *Delete* transition. Such deduction mechanisms are, in general, incomplete. If forward inferences are carried out in this method, their conclusions can only be treated as lemmas to be proved. (See, for example, the proof procedure of [27].) For instance, in Example 9, we have noted that the resolvent of the resolution inference completes the proof of case 4. In an implicit procedure, the resolvent must be treated as a lemma that again needs an inductive proof. This leads to wasteful computation, which can be even infinite. (Cf. Example 11.) While such instances of infinite wasteful computation may be rare, they underscore the inefficiencies caused by the lack of forward inferences in the deduction mechanism.

We can use the techniques of Section 6 to extend the implicit induction procedure with forward inferences:

PROCEDURE 5 (Explicit induction procedure for reductive inferences). The following transition rules act on pairs (P, H) where P is a set of propositions and H a set of supported propositions.

$$\begin{array}{ll}
\mathit{Expand} & (P \cup \{\phi\}, H) \Rightarrow (P \cup \Psi, H \cup \{\phi \rightrightarrows \phi\}) \\
 & \qquad\qquad\qquad \text{if } \Psi \text{ is a strict cover set for } \phi. \\
\mathit{Simplify} & (P \cup \{\phi\}, H) \Rightarrow (P \cup P', H) \\
 & \qquad\qquad\qquad \text{if } \mathcal{A} \cup P \cup P' \cup (H)_{\preccurlyeq\phi\rightrightarrows} \vdash^0_R \phi. \\
\mathit{Forward} & (P, H) \Rightarrow (P, H \cup \{S \rightrightarrows \phi\}) \\
 & \qquad\qquad\qquad \text{if } \mathcal{A} \cup H \vdash_{F_{\rightrightarrows}} (S \rightrightarrows \phi). \\
\mathit{Lemma} & (P, H) \Rightarrow (P \cup P', H).
\end{array}$$

This procedure is technically less general than Procedure 3, though we do not know whether this is significant in practice.

8. Conclusion

We have developed a concept of induction using syntactic orders defined over propositions. Such an extension offers a powerful deduction paradigm: the capability to do proof by mutual induction. With this technique, different theorems can share their inductive hypotheses; this often allows simpler or more direct proofs. Our examples illustrate the power of this technique and its ease of use. Formalizing and developing this paradigm is one of the contributions of this paper.

We have formalized the technique of implicit induction as a mechanism of induction using propositional orders. The essence of an implicit induction procedure is the use of reductive deduction relations and, implicit in such procedure, the use of proposition orders to control deductions. As a result, explicit induction tests to validate the use of the inductive hypotheses become unnecessary.

This continues the works of Hofbauer and Kutsche [18], Kounalis and Rusinowitch [25], and Reddy [31] to propose an interpretation of completion-based induction as classical Noetherian induction rather than the proof by consistency interpretation [22, 1].

The restriction to reductive deduction mechanisms, essential in implicit induction, limits the deductive capabilities of induction. We have described an effective induction procedure allowing both reductive and nonreductive inferences. This combines the advantages of implicit induction and explicit induction. This procedure is a proposed implementation of the technique of induction using propositional orders. It shows that with some simple bookkeeping mechanism one can divide the first-order deduction part of the procedure into combinations of reductive inferences and forward inferences. The resulting procedure has the simplicity and efficiency of implicit induction while lifting the restriction to reductive deductions.

Appendix

EXAMPLE OF MUTUAL INDUCTION

Consider the problem of proving the correctness of the unification algorithm given in Figure 1. The unification algorithm is defined for the following language: There are two parameters type Var and F whose elements denote variables and function symbols, respectively. There are three constructor types Term, TermList, and Subst. The type Term is defined with the constructors V: Var $\rightarrow$ Term and _[_]: F $\times$ TermList $\rightarrow$ Term, denoting variables and function applications. It satisfies the equation

$$\mathsf{Term} = V(\mathsf{Var}) \mid \mathsf{F}[\mathsf{TermList}].$$

The type TermList relies on the constructors Nil: TermList and _._: Term $\times$ TermList $\rightarrow$ TermList and satisfies the equation

$$\mathsf{TermList} = \mathsf{Nil} \mid \mathsf{Term}.\mathsf{TermList}.$$

The type Subst relies on the constructors Empty: Subst and $\{\_\mapsto\_\}.\_$: Var $\times$ Term $\times$ Subst $\rightarrow$ Subst and satisfies the equation

$$\mathsf{Subst} = \mathsf{Empty} \mid \{\mathsf{Var} \mapsto \mathsf{Term}\} \cdot \mathsf{Subst}.$$

The auxiliary functions involved in the theory are defined in Figure 2. Figure 3 shows some abbreviations that we use for readability.

We want to prove the correctness of the unification algorithm by induction, so we first need to define an appropriate order. Experience suggests that we should find a term order able to show the termination of the algorithm, and use that order in our inductive proof.

$$
\begin{array}{rcll}
& \mathsf{unify}: & \text{Returns the mgu of two unifiable terms} & \\
& \mathsf{unify}: & \mathsf{Term} \times \mathsf{Term} \to \mathsf{Subst} & \\
\mathsf{unify}(V(x), b) & \to & \textbf{if } V(x) \in \mathsf{vars}(b) \textbf{ then } \mathsf{Empty} \textbf{ else} & \\
& & \{x \mapsto b\} \cdot \mathsf{Empty} & (11) \\
\mathsf{unify}(a, V(x)) & \to & \textbf{if } V(x) \in \mathsf{vars}(a) \textbf{ then } \mathsf{Empty} \textbf{ else} & \\
& & \{x \mapsto a\} \cdot \mathsf{Empty} & (12) \\
\mathsf{unify}(f[l], g[m]) & \to & \textbf{if } f \neq g \textbf{ then } \mathsf{Empty} \textbf{ else } \mathsf{unify\text{-}l}(l, m) & (13) \\
& \mathsf{unify\text{-}l}: & \text{Returns the mgu of two unifiable lists of terms} & \\
& \mathsf{unify\text{-}l}: & \mathsf{TermList} \times \mathsf{TermList} \to \mathsf{Subst} & \\
\mathsf{unify\text{-}l}(\mathsf{Nil}, m) & \to & \mathsf{Empty} & (14) \\
\mathsf{unify\text{-}l}(l, \mathsf{Nil}) & \to & \mathsf{Empty} & (15) \\
\mathsf{unify\text{-}l}(a \cdot l, b \cdot m) & \to & \mathsf{unify}(a, b) \circ_s \ \mathsf{unify\text{-}l}(l \circ_l \ \mathsf{unify}(a, b), m \circ_l & \\
& & \mathsf{unify}(a, b)) & (16)
\end{array}
$$

Figure 1. Definition of the unification algorithm.

Showing the termination of the unification algorithm requires a sophisticated order. The difficulty comes from rule (16),

$$\mathsf{unify\text{-}l}(a \cdot l, b \cdot m) \to \mathsf{unify}(a, b) \circ_s \mathsf{unify\text{-}l}(l \circ_l \mathsf{unify}(a, b), m \circ_l \mathsf{unify}(a, b)).$$

To show that this rule is *decreasing* (i.e., that the left-hand side is greater than the right-hand side), we must, at the very least, show $\mathsf{unify\text{-}l}(a \cdot l, b \cdot m) \succ \mathsf{unify\text{-}l}(l \circ_l \mathsf{unify}(a, b), m \circ_l \mathsf{unify}(a, b))$. With most classical syntactic orders, this is impossible because this reduces to $a \cdot l \succ l \circ_l \mathsf{unify}(a, b)$ and $b \cdot m \succ m \circ_l \mathsf{unify}(a, b)$, and these cannot hold because of the free variables (respectively b and a) on the right side of the inequalities. However, these inequalities trivially hold if we use the semantic order in which lists are compared by comparing their length (keeping in mind that applying a substitution to a list does not change its length). Thus, when comparing the terms $\mathsf{unify\text{-}l}(a \cdot l, b \cdot m)$ and $\mathsf{unify\text{-}l}(l \circ_l \mathsf{unify}(a, b), m \circ_l \mathsf{unify}(a, b))$, we want to use this semantic relation as the complexity measure. However, to show that any of the other rules is decreasing, or even, in rule (16), that $\mathsf{unify\text{-}l}(a \cdot l, b \cdot m)$ is greater that $\mathsf{unify}(a, b)$, we need to use syntactic orders. Thus we need an order combining semantic and syntactic information.

We use the *general path order* of Dershowitz and Hoot [13] with five complexity measures, $\mu_0, \ldots, \mu_4$.

1. μ_0 maps terms to the set of function symbols.

$$\mu_0(f(\overrightarrow{m})) = f.$$

The well-founded order $>_p$ on the set of function symbols is

$$\text{unify} \approx_p \text{unify-l} > \circ_s > \circ_t \approx_p \circ_l > \text{vars} \approx_p \text{vars-l} > \text{embed}$$
$$\approx_p \text{topembed} > \in > \neq > \_ [\_] > \cdot > \text{Nil} > \{\_ \mapsto \_\}.\_ > \text{Empty}$$
$$> V > \text{“}\forall\text{”} \approx_p \textbf{ if } \approx_p \vee \approx_p \wedge$$

2. μ_1 maps terms to $\mathbb{N} + (\mathbb{N} \times \mathbb{N} \times \mathbb{N})$ where $+$ is the disjoint union (or discriminated union)

$$\mu_1(\text{unify}(a,b)) = (0, |\text{unify}(a,b)|)$$
$$\mu_1(\text{unify-l}(l,m)) = (1, (|\text{unify-l}(l,m)|, |l|, |m|))$$
$$\mu_1(f(\overrightarrow{m})) = (0,0) \qquad \text{if } f \neq \text{unify and } f \neq \text{unify-l}$$

Here, $|\text{unify}(a,b)|$ and $|\text{unify-l}(l,m)|$ are lengths of substitutions and $|l|$ is the length of lists.
The well-founded preorder $\geqslant_s$ on $\mathbb{N} + (\mathbb{N} \times \mathbb{N} \times \mathbb{N})$ is

$$(0,n) \geqslant_s (0,n') \Leftrightarrow n \geq n'$$
$$(1,(n,i,j)) \geqslant_s (1,(n',i',j')) \Leftrightarrow (n,i,j) \geqslant_{\text{lex}} (n',i',j')$$
$$(0,n) \geqslant_s (1,(n',i',j')) \Leftrightarrow n \geq n'$$
$$(1,(n,i,j)) \geqslant_s (0,n') \Leftrightarrow n > n'$$

3. μ_2 and μ_3 map terms to multisets of terms

$$\left.\begin{array}{l}\mu_2(t \circ \theta) = \{\theta\} \\ \mu_3(t \circ \theta) = \{t\}\end{array}\right\} \quad \text{for } \circ = \circ_t \text{ or } \circ = \circ_l$$
$$\left.\begin{array}{l}\mu_2(f(t_1,\ldots,t_n)) = \{t_1,\ldots,t_n\} \\ \mu_3(f(t_1,\ldots,t_n)) = \emptyset\end{array}\right\} \text{ for } f \neq \circ_t \text{ and } f \neq \circ_l$$

4. Finally, μ_4 maps terms to the set of function symbols, i.e., $\mu_4(f(\overrightarrow{m})) = f$. The well-founded order on the set of function symbols $>_{p'}$ is equal to $>_p$ except that embed $>$ topembed

The general path order with the above complexity measures takes the following form:

$$s = f(s_1,\ldots,s_n) \succ g(t_1,\ldots,t_m) = t$$

if either

1. $s_i \succcurlyeq t$ for some s_i, or
2. $s \succ t_1,\ldots,t_m$ and $(\mu_0(s),\ldots,\mu_4(s)) >_{\text{lex}} (\mu_0(t),\ldots,\mu_4(t))$ where $>_{\text{lex}}$ is the lexicographic combination of $>_p, >_s, \succ_{\text{mul}}, \succ_{\text{mul}}$; and $>_{p'}$ (where $\succ_{\text{mul}}$ is the multiset extension of $\succ$).

We note that $\succ$ satisfies the strict subterm property (i.e., $f(\ldots,s,\ldots) \succ s$) by the case 1 above. It is also stable (i.e., $s \succ t \Rightarrow s\sigma \succ t\sigma$). The order is not in

$\circ_t$: Apply a substitution to a term

$$\begin{aligned}
a \circ_t \mathsf{Empty} &\to a \\
V(x) \circ_t \{y \mapsto a\} \cdot \theta &\to \textbf{if } x = y \textbf{ then } a \textbf{ else } V(x) \circ_t \theta \\
f[l] \circ_t \theta &\to f[l \circ_l \theta]
\end{aligned}$$

$\circ_l$: Apply a substitution to a list of terms

$$\begin{aligned}
\mathsf{Nil} \circ_l \theta &\to \mathsf{Nil} \\
a \cdot l \circ_l \theta &\to (a \circ_t \theta) \circ (l \circ_l \theta)
\end{aligned}$$

$\circ_s$: Composition of two substitutions

$$\begin{aligned}
\mathsf{Empty} \circ_s \theta &\to \theta \\
\{y \mapsto a\} \cdot \theta_1 \circ_s \theta_2 &\to \{y \mapsto a \circ_t \theta_2\} \cdot (\theta_1 \circ_s \theta_2)
\end{aligned}$$

vars: Return the list of variables occurring in a term

$$\begin{aligned}
\mathsf{vars}(V(x)) &\to V(x) \cdot \mathsf{Nil} \\
\mathsf{vars}(f[l]) &\to \mathsf{vars\text{-}l}(l)
\end{aligned}$$

vars-l: Return the list of variables occurring in a list of terms

$$\begin{aligned}
\mathsf{vars\text{-}l}(\mathsf{Nil}) &\to \mathsf{Nil} \\
\mathsf{vars\text{-}l}(a \cdot l) &\to \mathsf{vars}(a) + +\mathsf{vars\text{-}l}(l)
\end{aligned}$$

$\in$: List membership relation

$$\begin{aligned}
x \in \mathsf{Nil} &\to \mathsf{False} \\
x \in a \cdot l &\to x = a \vee x \in l
\end{aligned}$$

$++$: List concatenation

$$\begin{aligned}
\mathsf{Nil}{+}{+}l &\to l \\
a \cdot l{+}{+}m &\to a \cdot (l{+}{+}m)
\end{aligned}$$

embed: Strict embedding relation:

embed: (Term ∪ TermList) × (Term ∪ TermList) → Boolean

$$\begin{aligned}
\mathsf{embed}(t, V(x)) &\to \mathsf{False} \\
\mathsf{embed}(t, f[m]) &\to t = m \vee \mathsf{embed}(t, m) \vee \mathsf{topembed}(t, f[m]) \\
\mathsf{embed}(t, \mathsf{Nil}) &\to \mathsf{False} \\
\mathsf{embed}(t, a \cdot m) &\to t = a \vee \mathsf{embed}(t, a) \vee t = m \vee \\
&\quad\ \mathsf{embed}(t, m) \vee \mathsf{topembed}(t, a \cdot m)
\end{aligned}$$

topembed: Strict embedding relation with match at the top position

$$\begin{aligned}
\mathsf{topembed}(V(x), b) &\to \mathsf{False} \\
\mathsf{topembed}(f[l], g[m]) &\to f = g \wedge \mathsf{embed}(l, m) \\
\mathsf{topembed}(\mathsf{Nil}, b) &\to \mathsf{False} \\
\mathsf{topembed}(a \cdot l, g[m]) &\to \mathsf{False} \\
\mathsf{topembed}(g[l], a \cdot x) &\to \mathsf{False} \\
\mathsf{topembed}(a \cdot l, b \cdot m) &\to \mathsf{embed}(a, b) \wedge (l = m \vee \mathsf{embed}(l, m)) \vee \\
&\quad\ ((a = b \vee \mathsf{embed}(a, b)) \wedge \mathsf{embed}(l, m))
\end{aligned}$$

Figure 2. Auxiliary functions for the unification algorithm.

$$\text{unifiable}(a,b) \stackrel{\text{def}}{=} \exists\theta, a \circ_t \theta = b \circ_t \theta$$
$$\text{unifiable-l}(l,m) \stackrel{\text{def}}{=} \exists\theta, l \circ_l \theta = m \circ_t \theta$$
$$\theta_2 \geq \theta_1 \stackrel{\text{def}}{=} \forall l, l \circ_t \theta_2 = l \circ_t \theta_1 \circ_t \theta_2$$

Figure 3. Abbreviations.

general *strictly* monotonic, i.e., $s \succ t$ does not imply $u[s] \succ u[t]$, though, for appropriate contexts u, the property holds.

Using the order $\succ$, we can show the termination of the unification algorithm, i.e., all the rules given in Figure 1 are decreasing. To show that rule (16) is decreasing, we note that unify-l $>_p \circ_s$. Thus, we must show

$$\text{unify-l}(a \cdot l, b \cdot m) \succ \text{unify}(a,b)$$
$$\text{unify-l}(a \cdot l, b \cdot m) \succ \text{unify-l}(l \circ_l \text{unify}(a,b), m \circ_l \text{unify}(a,b))$$

The first inequality may be seen by noting

$$\mu_1(\text{unify-l}(a \cdot l, b \cdot m)) \geqslant_s \mu_1(\text{unify}(a,b)) \text{ and}$$
$$\mu_2(\text{unify-l}(a \cdot l, b \cdot m)) \succ_{\text{mul}} \mu_2(\text{unify}(a,b))$$

The second inequality follows from

$$\mu_1(\text{unify-l}(a \cdot l, b \cdot m)) >_s \mu_1(\text{unify-l}(l \circ_l \text{unify}(a,b), m \circ_l \text{unify}(a,b)))$$

In showing that rule (13) is decreasing, the main facts involved are

$$\mu_1(\text{unify}(f[l], g[m])) \geqslant_s \mu_1(\text{unify-l}(l,m))$$
$$\mu_2(\text{unify}(f[l], g[m])) \succ_{\text{mul}} \mu_2(\text{unify-l}(l,m))$$

Having defined the order $\succ$, we are ready to prove the correctness of the unification algorithm.

THEOREM 13. *For all terms a and b, if a and b are unifiable then* unify(a,b) *computes a most general unifier of a and b. Formally,*

$$\forall a,b\colon \text{Term}.P(a,b)$$

where $P(a,b)$ is

$$P(a,b)\colon \text{unifiable}(a,b) \Rightarrow a \circ_t \text{unify}(a,b) = b \circ_t \text{unify}(a,b) \wedge$$
$$\forall\theta \cdot a \circ_t \theta = b \circ_l \theta \Rightarrow \theta \geqslant \text{unify}(a,b) \qquad (17)$$

Proof. We select the cover set of substitutions $\{a \to V(x)\}, \{b \to V(x)\}$, and $\{a \to f[l], b \to g[m]\}$.

Case $a \to V(x)$: Proposition 17 simplifies, using the rules given in Figures 1 and 2, to

$$V(x) \notin \mathsf{vars}(b) \Rightarrow b \circ_t \{x \mapsto t\} = b \tag{18}$$
$$V(x) \notin \mathsf{vars\text{-}l}(l) \Rightarrow l \circ_l \{x \mapsto t\} = l^* \tag{19}$$
$$V(x) \circ_t \theta = b \circ_t \theta \Rightarrow \theta \geqslant \{x \mapsto b\} \tag{20}$$
$$\neg\mathsf{embed}(V(x), b) \wedge V(x) \in \mathsf{vars}(b) \Rightarrow V(x) = b \tag{21}$$
$$V(x) \in \mathsf{vars\text{-}l}(l) \Rightarrow \mathsf{embed}(V(x), l)^* \tag{22}$$
$$\neg\mathsf{embed}(x, x) \tag{23}$$
$$\mathsf{embed}(x, y) \Rightarrow \neg\mathsf{embed}(y, x) \tag{24}$$
$$\mathsf{embed}(a \cdot x, y) \Rightarrow \mathsf{embed}(a, y) \wedge \mathsf{embed}(x, y) \tag{25}$$
$$\mathsf{embed}(x, y) \Rightarrow \mathsf{embed}(x \circ_t \theta, y \circ_t \theta) \tag{26}$$
$$\mathsf{embed}(x, y) \Rightarrow \mathsf{embed}(x \circ_l \theta, y \circ_l \theta) \tag{27}$$
$$\mathsf{embed}(x, y) \Rightarrow \mathsf{embed}(x \circ_t \theta, y \circ_l \theta) \tag{28}$$
$$\mathsf{embed}(x, y) \Rightarrow \mathsf{embed}(x \circ_l \theta, y \circ_t \theta) \tag{29}$$
$$a \circ_t (\theta_1 \circ_s \theta_2) \rightarrow a \circ_t \theta_1 \circ_t \theta_2 \tag{30}$$
$$l \circ_l (\theta_1 \circ_s \theta_2) \rightarrow l \circ_l \theta_1 \circ_l \theta_2 \tag{31}$$
$$l \circ_l \theta = \mathsf{Nil} \Rightarrow l = \mathsf{Nil} \tag{32}$$
$$\mathsf{unifiable\text{-}l}(a \cdot l, b \cdot m) \Rightarrow \mathsf{unifiable}(a, b) \wedge \mathsf{unifiable\text{-}l}(l, m) \tag{33}$$
$$(\forall a \cdot a \circ_t \theta = a \circ_t \theta_1 \circ_t \theta) \Rightarrow l \circ_l \theta = l \circ_l \theta_1 \circ_l \theta \tag{34}$$

Figure 4. Lemmas used in the proof of the correctness of the unification algorithm. The lemmas noted by asterisks are automatically generated by the induction procedure.

$$\begin{aligned} \mathsf{unifiable}(V(x), b) \Rightarrow\ & \textbf{if } V(x) \in \mathsf{vars}(b) \\ & \textbf{then } V(x) = b \\ & \textbf{else } b = b \circ_t \{x \mapsto b\} \wedge \\ & \quad (\forall \theta \cdot V(x) \circ_t \theta = b \circ_t \theta \Rightarrow \theta \geqslant \{x \mapsto b\}) \end{aligned} \tag{35}$$

If we use Lemmas 18 and 20 of Figure 4, the else-branch of this proposition simplifies to True. To show that the then-branch is also true, we can use Lemma (21) if we can show

$$\mathsf{unifiable}(V(x), b) \Rightarrow \neg\mathsf{embed}(V(x), b) \tag{36}$$

This, in turn, follows from the definition of $\mathsf{unifiable}(V(x), b)$ and Lemmas 26 and 23. More precisely, the definition of $\mathsf{unifiable}(V(x), b)$ and Lemma 26 entails

$$\begin{aligned} \mathsf{unifiable}(V(x), b) \Rightarrow\ & \exists\theta, V(x) \circ_t \theta \\ & = b \circ_t \theta \wedge (\neg\mathsf{embed}(V(x), b) \vee \mathsf{embed}(x \circ_t \theta, x \circ_t \theta)) \end{aligned}$$

Using Lemma 23 we simplify this to the desired Proposition 36 above.

Case $b \to V(x)$: This case is symmetric to the previous one.

Case $a \to f[l], b \to g[m]$: Proposition 17 simplifies to

$$Q(l,m)\text{: } \mathsf{unifiable\text{-}l}(l,m) \Rightarrow l \circ_l \mathsf{unify\text{-}l}(l,m) = m \circ_l \mathsf{unify\text{-}l}(l,m) \wedge \\ \forall\theta \cdot l \circ_l \theta = m \circ_l \theta \Rightarrow \theta \geqslant \mathsf{unify\text{-}l}(l,m) \quad (37)$$

We cannot prove this proposition using first-order reasoning. Thus, we try to prove it by induction. Herein lies the power of mutual induction: this unresolved case can be treated as a lemma needing an inductive proof and, in that proof, we can use the inductive hypothesis of Theorem 13.

Let $Q(l,m)$ denote Proposition 37. It is proved as follows: We select the cover set $\{l \to \mathsf{Nil}\}, \{m \to \mathsf{Nil}\}$, and $\{l \to a \cdot l', m \to b \cdot m'\}$.

Cases $l \to$ Nil *and* $m \to$ Nil: Both these cases simplify to True using the rules given in Figures 1 and 2.

Case $l \to a \cdot l', m \to b \cdot m'$: After simplification using the rules given in Figures 1 and 2 and Lemmas 30 and 31 from Figure 4, $Q(a \cdot l', b \cdot m')$ becomes the following: Let σ_1 denote $\mathsf{unify}(a,b)$, and σ_2 denote $\mathsf{unify\text{-}l}(l \circ_l \sigma_1, m \circ_l \sigma_1)$, the simplified proposition is

$$\mathsf{unifiable\text{-}l}(a \cdot l', b \cdot m') \Rightarrow \\ a \circ_t \sigma_1 \circ_t \sigma_2 = b \circ_t \sigma_1 \circ_t \sigma_2 \wedge l' \circ_l \sigma_1 \circ_l \sigma_2 = m' \circ_l \sigma_1 \circ_l \sigma_2 \wedge \\ \forall\theta \cdot a \circ_t \theta = b \circ_t \theta \wedge l' \circ_l \theta = m' \circ_l \theta \Rightarrow \theta \geqslant \sigma_1 \circ_s \sigma_2 \quad (38)$$

We will prove, in turn, that the antecedent of Proposition 38 implies each conjunct in the consequent.

- $a \circ_t \mathsf{unify}(a,b) \circ_t \mathsf{unify\text{-}l}(l \circ_l \mathsf{unify}(a,b), m \circ_l \mathsf{unify}(a,b)) = b \circ_t \mathsf{unify}(a,b) \circ_t \mathsf{unify\text{-}l}(l \circ_l \mathsf{unify}(a,b), m \circ_l \mathsf{unify}(a,b))$: Proving this equality requires the inductive hypothesis of Theorem 13. From the antecedent $\mathsf{unifiable\text{-}l}(a \cdot l', b \cdot m')$ and Lemma 33, we conclude $\mathsf{unifiable}(a,b)$. Thus, the instance $P(a,b)$ of Theorem 13 is applicable to show $a \circ_t \mathsf{unify}(a,b) = b \circ_t \mathsf{unify}(a,b)$. The use of $P(a,b)$ is valid since $P(a,b)$ is smaller than $Q(a \cdot l', b \cdot m')$ by the order $\succ$.
- $l' \circ_l \mathsf{unify}(a,b) \circ_l \mathsf{unify\text{-}l}(l \circ_l \mathsf{unify}(a,b), m \circ_l \mathsf{unify}(a,b)) = m' \circ_l \mathsf{unify}(a,b) \circ_l \mathsf{unify\text{-}l}(l \circ_l \mathsf{unify}(a,b), m \circ_l \mathsf{unify}(a,b))$: This requires the inductive hypotheses of both Theorem 13 and Proposition 37. As shown above, $\mathsf{unifiable}(a,b)$ holds. Thus, the inductive hypothesis $P(a,b)$ implies

  $$a \circ_t \theta = b \circ_t \theta \Rightarrow l \circ_t \theta = l \circ_t \mathsf{unify}(a,b) \circ_t \theta$$

  From Lemma 34, this implies

$$a \circ_t \theta = b \circ_t \theta \Rightarrow l' \circ_l \theta = l' \circ_l \mathsf{unify}(a, b) \circ_l \theta \tag{39}$$

From the antecedent $\mathsf{unifiable\text{-}l}(a \cdot l', b \cdot m')$ we derive

$$\exists \theta, a \circ_t \theta = b \circ_t \theta \wedge l' \circ_l \theta = m' \circ_l \theta$$

With Proposition 39 this implies $\exists \theta, l' \circ_l \mathsf{unify}(a, b) \circ_l \theta = m' \circ_l \mathsf{unify}(a, b) \circ_l \theta$. Thus, $\mathsf{unifiable\text{-}l}(l' \circ_l \mathsf{unify}(a, b), m' \circ_l \mathsf{unify}(a, b))$ holds, and we can use the instance $Q(l' \circ_l \mathsf{unify}(a, b), m' \circ_l \mathsf{unify}(a, b))$ of the inductive hypothesis of Proposition 37 to complete the proof of this case.
The use of $Q(l' \circ_l \mathsf{unify}(a, b), m' \circ_l \mathsf{unify}(a, b))$ is valid, since $Q(l' \circ_l \mathsf{unify}(a, b), m' \circ_l \mathsf{unify}(a, b))$ is smaller than $Q(a \cdot l', b \cdot m')$ by the order $\succ$.

– $\forall \theta \cdot a \circ_t \theta = b \circ_t \theta \wedge l' \circ_l \theta = m' \circ_l \theta \Rightarrow \theta \geqslant \mathsf{unify}(a, b) \circ_s \mathsf{unidy\text{-}l}(l \circ_l \mathsf{unify}(a, b), m \circ_l \mathsf{unify}(a, b))$: After simplification by Lemma 30, we are left to show

$$\begin{aligned} &\forall \theta \cdot a \circ_t \theta = b \circ_t \theta \wedge l' \circ_l \theta = m' \circ_l \theta \Rightarrow \\ &\quad \forall l \cdot l \circ_t \theta = l \circ_t \mathsf{unify}(a, b) \\ &\qquad\qquad \circ_t \mathsf{unify\text{-}l}(l \circ_l \mathsf{unify}(a, b), m \circ_l \mathsf{unify}(a, b)) \circ_t \theta \end{aligned} \tag{40}$$

As shown in the previous cases, $\mathsf{unifiable}(a, b)$ and $\mathsf{unifiable\text{-}l}(l' \circ_l \mathsf{unify}(a, b), m' \circ_l \mathsf{unify}(a, b))$ hold. Thus, the inductive hypothesis $Q(l' \circ_l \mathsf{unify}(a, b), m' \circ_l \mathsf{unify}(a, b))$ implies

$$l' \circ_l \mathsf{unify}(a, b) \circ_l \theta = m' \circ_l \mathsf{unify}(a, b) \circ_l \theta \Rightarrow \forall l' \cdot l' \circ_t \theta = l' \circ_t \sigma_2 \circ_t \theta$$

while $P(a, b)$ implies

$$a \circ_t \theta = b \circ_t \theta \Rightarrow \forall l' \cdot l' \circ_t \theta = l' \circ_t \mathsf{unify}(a, b) \circ_t \theta$$

Proposition 40 follows from these.
This completes the proof of Proposition 37, which in turn complete the proof of Theorem 13.

□

Automation of the proof. We machine-checked the larger part of the proof presented above in our experimental Focus system. The system could check all the deductive inferences with minimal user intervention. However, the well-foundedness of the term order and the validity of the use of the inductive hypotheses had to be certified by the user.

The deductive capabilities of Focus include implicit induction, explicit induction with term orders that are instances of the recursive path ordering, and a strong first-order deduction component. The inductive component of Focus implements the Explicit induction procedure given earlier. The first-order deduction component combines a reductive deduction mechanism for automatic simplifications and a user-controlled completion mechanism allowing forward inferences.

We discussed our induction method in the body of this paper. Our first-order deduction mechanism is discussed in detail in [7–9]. Let us simply say here that the completion mechanism uses resolution and paramodulation, while the reductive deduction mechanism relies on hyperresolution (multiple resolution inferences combined in one step) and paramodulation. The reductive deduction mechanism insists that the result of an inference be strictly smaller than the largest goal premise. This restriction ensures that this is indeed a reductive mechanism. Further, it guarantees that deductions always terminate. Hence, simplification can be automatic. We illustrate these deduction mechanisms with the following examples:

1. The following proposition is derived from the inductive hypothesis $P(a, b)$:

   $$\mathsf{unifiable}(a, b) \Rightarrow (\forall\theta \cdot a \circ_t \theta = b \circ_t \theta \Rightarrow \theta \geq \mathsf{unify}(a, b))$$

   Simplifying unifiable and $\geq$, we obtain

   $$\begin{aligned}&(\exists\theta, a \circ_t \theta = b \circ_t \theta) \wedge a \circ_t \theta_2 = b \circ_t \theta_2\\ &\quad\Rightarrow \forall l \cdot l \circ_t \theta_2 = l \circ_t \mathsf{unify}(a, b) \circ_t \theta_2\end{aligned}$$

   The completion mechanism can resolve this proposition and Lemma 34 to produce

   $$\begin{aligned}\exists\theta, a \circ_t \theta &= b \circ_t \theta \wedge a \circ_t \theta_2 = b \circ_t \theta_2\\ &\Rightarrow \forall l \cdot l \circ_l \theta_2 = l \circ_l \mathsf{unify}(a, b) \circ_l \theta_2\end{aligned} \tag{41}$$

   Since Proposition 41 is derived from the inductive hypothesis $P(a, b)$, it has the support $\{P(a, b)\}$.

2. Consider the statement made in the proof of the second conjunct of Proposition 38

   "From the antecedent unifiable-l$(a \cdot l', b \cdot m')$ we derive

   $$\exists\theta, a \circ_t \theta = b \circ_t \theta \wedge l' \circ_l \theta = m' \circ_l \theta$$

   With Proposition 39 this implies $\exists\theta, l' \circ_l \mathsf{unify}(a, b) \circ_l \theta = m' \circ_l \mathsf{unify}(a, b) \circ_l \theta$. Thus, unifiable-l$(l' \circ_l \mathsf{unify}(a, b), m' \circ_l \mathsf{unify}(a, b))$ holds, and we can use the instance $Q(l' \circ_l \mathsf{unify}(a, b), m' \circ_l \mathsf{unify}(a, b)) \ldots$."

   In the machine proof, instead of this reasoning, we rely on the completion mechanism as follows: From the inductive hypothesis $Q(l, m)$ we construct the proposition

   $$\exists\theta, l \circ_l \theta = m \circ_l \theta \Rightarrow l \circ_l \mathsf{unify\text{-}l}(l, m) = m \circ_l \mathsf{unify\text{-}l}(l, m)$$

   Using a paramodulation inference, the completion mechanism combines the above proposition and Proposition 41 to produce (σ_1 denotes unify(a, b))

   $$\begin{aligned}&(\exists\theta, a \circ_t \theta = b \circ_t \theta) \wedge (\exists\theta_2, a \circ_t \theta_2 = b \circ_t \theta_2 \wedge l \circ_l \theta_2 = m \circ_l \theta_2) \Rightarrow\\ &\quad l \circ_l \sigma_1 \circ_l \mathsf{unify\text{-}l}(l \circ_l \sigma_1, m \circ_l \sigma_1)\\ &\quad = m \circ_l \sigma_1 \circ_l \mathsf{unify\text{-}l}(l \circ_l \sigma_1, m \circ_l \sigma_1)\end{aligned} \tag{42}$$

This proposition will carry the support

$$\{P(a,b), Q(l \circ_l \mathsf{unify}(a,b), m \circ_l \mathsf{unify}(a,b))\}$$

since it is derived from both the inductive hypotheses.
Using the derived rule (42), the reductive deduction mechanism can easily complete the proof of the second conjunct of Proposition 38 as shown below.

3. Consider the proof of the second conjunct of Proposition 38. We want to prove

$$\mathsf{unifiable\text{-}l}(a \cdot l, b \cdot m) \Rightarrow l \circ_l \sigma_1 \circ_l \sigma_2 = m \circ_l \sigma_1 \circ_l \sigma_2$$

The goal is skolemized and negated and becomes (θ_2 is a Skolem function of a and b)

$$a \cdot l \circ_l \theta_2 = b \cdot m \circ_l \theta_2 \wedge l \circ_l \sigma_1 \circ_l \sigma_2 \neq m \circ_l \sigma_1 \circ_l \sigma_2$$

This proof is

$$\dfrac{\dfrac{\dfrac{a \cdot l \circ_l \theta_2 = b \cdot m \circ_l \theta_2}{(a \circ_t \theta_2) \cdot (l \circ_l \theta_2) = (b \circ_t \theta_2) \cdot (m \circ_l \theta_2)}\,\text{Def. of } \circ_l}{a \circ_t \theta_2 = b \circ_t \theta_2 \wedge l \circ_l \theta_2 = m \circ_l \theta_2} \qquad l \circ_l \sigma_1 \circ_l \sigma_2 \neq m \circ_l \sigma_1 \circ_l \sigma_2}{\bot}\,(42)$$

4. To prove Proposition 40, we proceed as follows: The completion mechanism is used to derive

$$(\exists \theta, a \circ_t \theta = b \circ_t \theta) \wedge (\exists \theta_2, a \circ_t \theta_2 = b \circ_t \theta_2 \wedge l' \circ_l \theta_2 = m' \circ_l \theta_2) \Rightarrow$$
$$((l \circ_l \sigma_1) \circ_l \theta_3 = (m \circ_l \sigma_1) \circ_l \theta_3 \Rightarrow \forall l' \cdot l' \circ_t \theta_3 = l' \circ_t \sigma_2 \circ_t \theta_3) \quad (43)$$

in the same manner as Proposition 42 was derived. The goal, skolemized and negated, produces (where θ and l are Skolem constants)

$$a \circ_t \theta = b \circ_t \theta \quad (44)$$
$$l' \circ_l \theta = m' \circ_l \theta \quad (45)$$
$$l \circ_t \theta \neq l \circ_t \sigma_1 \circ_t \sigma_2 \circ_t \theta \quad (46)$$

The precedence order is extended with $l >_p l >_p m$, and we obtain the proof

$$\dfrac{\dfrac{\dfrac{(44) \quad (44) \quad (45) \quad (46)}{(l' \circ_l \sigma_1) \circ_l \theta \neq (m' \circ_l \sigma_1) \circ_l \theta}\,(43) \qquad (44)}{l' \circ_l \theta \neq m' \circ_l \theta}\,(41) \qquad (45)}{\bot}$$

Comparison with Rouyer and Lescanne's proof using Coq. Rouyer and Lescanne studied the unification problem with the Coq proof assistant [35]. The differences between their proof and ours are summarized in Figure 5. However, we should

	Our proof	Rouyer and Lescanne's Proof	
		Total	Unification only
Theorems and lemmas	17	48	36
Definitions	14	43	28
Length of proof script	15 pages	48 pages	35 pages

Figure 5. Comparison between our and Rouyer and Lescanne's proof.

recall that Rouyer and Lescanne's proof is more thorough than ours, so we included in the fourth column an estimation of the size of their proof restricted to the unification problem.

Interestingly, the difference in complexity between our proof and Rouyer and Lescanne's proof seems to arise from the use of conventional induction. The correctness theorem has the form

$$\forall x\colon\ \tau_1 \cdot \phi(x) \wedge \forall y\colon\ \tau_2 \cdot \psi(y)$$

where τ_1 is the type of terms, τ_2 is the type of lists of terms, $\phi(x)$ expresses the correctness of the unify function for terms, and $\psi(y)$ expresses the correctness of the unify function for lists of terms. The difficulty is that the properties $\phi(x)$ and $\psi(y)$ need each other in their proofs in a mutually recursive fashion. This is not a problem in our approach due to the use of term orders. However, it is unclear how to use conventional induction for such a theorem. Rouyer and Lescanne's proof relied on the introduction of the type of *quasi-terms* to combine both terms and lists of terms into one type. Then, they solved the unification problem for quasi-terms and showed that this implied that the unification problem for terms was solved as well. This introduction of a new type resulted in a large increase in the number of definitions.

Acknowledgements

We thank Sergeï G. Vorobyov for many valuable discussions and explanations, and Adel Bouhoula, Nachum Dershowitz, Michaël Rusinowitch, and Pierre Lescanne for fruitful discussions.

References

1. Bachmair, L.: Proof by consistency in equational theories, in *Proc. 3rd LICS Symp.*, 1988, pp. 228–233.
2. Bachmair, L., Dershowitz, N., and Plaisted, D. A.: Completion without failure, in H. Aït-Kaci and M. Nivat (eds), *Resolutions of Equations in Algebraic Structures*, Vol. 2: Rewriting Techniques, Chapter 1, Academic Press, New York, 1989, pp. 1–30.
3. Biundo, S., Hummel, B., Hutter, D., and Walther, C.: The Karlsruhe induction theorem proving system, in *8th CADE Conf.*, LNCS 230, Springer-Verlag, 1986.

4. Bouhoula, A. and Rusinowitch, M.: Automatic case analysis in proof by induction, in R. Bajcsy (ed.), *Proc. 13th IJCAI Conf.*, Vol. 1, Morgan Kaufmann, August 1993, pp. 88–94.
5. Boyer, R. S. and Moore, J. S.: *A Computational Logic*, Academic Press, New York, 1979.
6. Boyer, R. S. and Moore, J. S.: A theorem prover for a computational logic, in M. E. Stickel (ed.), *Proc. 10th CADE Conf.*, LNCS 449, Springer-Verlag, 1990, pp. 1–15.
7. Bronsard, F.: Using term ordering to control clausal deductions, PhD thesis, University of Illinois, Urbana, 1995.
8. Bronsard, F. and Reddy, U. S.: Conditional rewriting in Focus, in S. Kaplan and M. Okada (eds), *Proc. 2nd CTRS Workshop*, LNCS 516, Springer-Verlag, 1991, pp. 2–13.
9. Bronsard, F. and Reddy, U. S.: Reduction techniques for first-order reasoning, in M. Rusinowitch and J. L. Rémy (eds), *Proc. 3rd CTRS Workshop*, LNCS 656, Springer-Verlag, 1992, pp. 242–256.
10. Bronsard, F., Reddy, U. S., and Hasker, R. W.: Induction using term orderings, in Alan Bundy (ed.), *Proc. 12th CADE Conf.*, LNAI 814, Springer-Verlag, 1994, pp. 102–117.
11. Bundy, A.: A rational reconstruction and extension of recursion analysis, in *IJCAI*, 1989.
12. Dershowitz, N.: Completion and its applications, in *Resolution of Equations in Algebraic Structures*, Vol. 2: Rewriting Techniques, Academic Press, San Diego, 1989, pp. 31–86.
13. Dershowitz, N. and Hoot, C.: Natural termination, in C. Kirchner (ed.), *Proc. 5th RTA Conf.*, LNCS 690, Montreal (Canada), Springer-Verlag, 1993, pp. 405–420.
14. Dershowitz, N. and Jouannaud, J.-P.: Rewrite systems, in J. van Leeuwen (ed.), *Handbook of Theoretical Computer Science B: Formal Methods and Semantics*, Chapter 6, North-Holland, Amsterdam, 1990, pp. 243–320.
15. Fribourg, L.: A strong restriction of the inductive completion procedure, *J. Symbolic Computation*, **8**(3) (1989), 253–276.
16. Goguen, J. A.: How to prove inductive hypotheses without induction, in *5th CADE Conf.*, LNCS 87, Springer-Verlag, 1980, pp. 356–372.
17. Gramlich, B.: Induction theorem proving using refined unfailing completion techniques, in *ECAI*, 1989 (also Technical Report SR-89-14, Universität Kaiserslautern, Germany).
18. Hofbauer, D. and Kutsche, R. D.: Proving inductive theorems based on term rewriting systems, in J. Grabowski, P. Lescanne, and W. Wechler (eds), *Proc. 1st ALP Workshop*, Akademie Verlag, 1988, pp. 180–190.
19. Hsiang, J. and Rusinowitch, M.: On word problems in equational theories, in T. Ottmann (ed.), *14th Intern. Colloq. Automata, Languages and Programming*, LNCS 267, Springer-Verlag, July 1987, pp. 54–71.
20. Huet, G. and Hullot, J.-M.: Proofs by induction in equational theories with constructors, *J. of Computer and System Sciences* **25** (1982), 239–266.
21. Jouannaud, J.-P. and Kounalis, E.: Automatic proofs by induction in equational theories without constructors, *Information and Computation* **82** (1989), 1–33. Original version in Symp. LICS, IEEE, 1986.
22. Kapur, D. and Musser, D. R.: Proof by consistency, *Artificial Intelligence* **31**(2) (Feb. 1987), 125–157.
23. Kapur, D., Narendran, P., and Zhang, H.: Automating inductionless induction using test sets, *J. Symbolic Computation* **11** (1991), 83–112.
24. Knuth, D. and Bendix, P.: Simple word problems in universal algebras, in J. Leech (ed.), *Computational Problems in Abstract Algebra*, Pergamon, Oxford, 1970, pp. 263–297.
25. Kounalis, E. and Rusinowitch, M.: Mechanizing inductive reasoning, in *Proc. AAAI Conf.*, AAAI Press and MIT Press, Boston, 1990, pp. 240–245.
26. Kowalski, R.: Studies in the completeness and efficiency of theorem-proving by resolution, PhD thesis, University of Edinburgh, 1970.
27. Küchlin, W.: Inductive completion by ground proof transformation, in H. Aït-Kaci and M. Nivat (eds), *Resolution of Equations in Algebraic Structures*, Vol. 2: Rewriting Techniques, Academic Press, San Diego, 1989, pp. 211–245.
28. Lankford, D. S.: A simple explanation of inductionless induction, Memo MTP-14, Dept. of Mathematics, Louisiana Tech. Univ., August 1981.

29. McAllester, D.: Term rewriting induction, `theorem - provers@ai.mit.edu` electronic bulletin board, 1990.
30. Musser, D. R.: On proving inductive properties of abstract data types, in *ACM Symp. on Princ. of Program. Lang.*, ACM, 1980, pp. 154–162.
31. Reddy, U. S.: Term rewriting induction, in M. Stickel (ed.), *10th CADE Conf.*, LNAI 449, Springer-Verlag, 1990, pp. 162–177.
32. Rémy, J.-L.: *Etude des Systèmes de Réécriture Conditionnels et Applications aux Types Abstraits Algébriques*, Th. Etat, INPL, Nancy (France), 1982.
33. Robinson, G. A. and Wos, L. T.: Paramodulation and first-order theorem proving, in B. Meltzer and D. Mitchie (eds), *Machine Intelligence 4*, Edinburgh University Press, 1969, pp. 135–150.
34. Robinson, J. A.: A machine-oriented logic based on the resolution principle, *JACM* **12** (1965), 12–23.
35. Rouyer, J. and Lescanne, P.: Verification and programming of first-order unification in the calculus of constructions with inductive types, November 1992.
36. Selman, A.: Completeness of calculi for axiomatically defined classes of algebras, *Algebra Universalis* **2** (1972), 20–32.
37. Wirth, C.-P. and Gramlich, B.: On notions of inductive validity for first-order equational clauses, in *12th CADE Conf.,* LNCS 814, Springer-Verlag, 1994, pp. 162–176.
38. Zhang, H. and Kapur, D., and Krishnamoorthy, M. S.: A mechanizable induction principle for equational specifications, in E. Lusk and R. Overbeek (eds), *9th CADE Conf.*, LNCS 310, Springer-Verlag, 1988, pp. 162–181.

Journal of Automated Reasoning **16**: 39–78, 1996.

New Uses of Linear Arithmetic in Automated Theorem Proving by Induction*

DEEPAK KAPUR and M. SUBRAMANIAM
Computer Science Department, State University of New York, Albany, NY 12222, U.S.A.
e-mail: kapur@cs.albany.edu, subu@cs.albany.edu

Abstract. Zhang, Kapur, and Krishnamoorthy introduced a cover set method for designing induction schemes for automating proofs by induction from specifications expressed as equations and conditional equations. This method has been implemented in the theorem prover *Rewrite Rule Laboratory (RRL)* and a proof management system *Tecton* built on top of *RRL*, and it has been used to prove many nontrivial theorems and reason about sequential as well as parallel programs. The cover set method is based on the assumption that a function symbol is defined by using a finite set of terminating (conditional or unconditional) rewrite rules. The termination ordering employed in orienting the rules is used to perform proofs by well-founded induction. The left sides of the rules are used to design different cases of an induction scheme, and recursive calls to the function made in the right side can be used to design appropriate instantiations for generating induction hypotheses. A weakness of this method is that it relies on syntactic unification for generating an induction scheme for a conjecture. This paper goes a step further by proposing semantic analysis for generating an induction scheme for a conjecture from a cover set. We discuss the use of a decision procedure for Presburger arithmetic (quantifier-free theory of numbers with the addition operation and relational predicates $>$, $<$, $\neq$, $=$, $\geqslant$, $\leqslant$) for performing semantic analysis about numbers. The decision procedure is used to generate appropriate induction schemes for a conjecture by using cover sets of function taking numbers as arguments. This extension of the cover set method automates proofs of many theorems that otherwise require human guidance and hints. The effectiveness of the method is demonstrated by using some examples that commonly arise in reasoning about specifications and programs. It is also shown how semantic analysis using a Presburger arithmetic decision procedure can be used for checking the completeness of a cover set of a function defined by using operations such as $+$ and $-$ on numbers. With this check, many function definitions used in a proof of the prime factorization theorem stating that every number can be factored uniquely into prime factors, which had to be checked manually, can now be checked automatically in *RRL*. The use of the decision procedure for guiding generalization for generating conjectures and merging induction schemes is also illustrated.

Key words: Induction, automated theorem proving, heuristics, linear arithmetic, Presburger arithmetic, generalization, semantic unification.

1. Introduction

While mechanizing proofs by induction, a crucial problem that often needs to be addressed is to decide on an induction scheme that leads to an appropriate induction hypothesis (or a set of induction hypotheses) to carry out the proof. This is related to an appropriate choice of a variable (or a finite set of vari-

* Partially supported by the National Science Foundation Grant No. CCR-9303394 and subcontract CB0249 of SRI contract MDA904-92-C-5186 with The Maryland Procurement Office.

ables) in a conjecture to perform induction on. In their seminal work, Boyer and Moore proposed designing induction schemes based on terminating function definitions. Typically the definitions of function symbols appearing in a conjecture are considered as candidates for designing induction schemes. The soundness of an induction scheme so designed is based on a well-founded ordering used to show the termination of the function definition. This insight often results in generation of appropriate induction hypotheses that, in many cases, can be used to prove (or disprove) a conjecture. This approach is supported in Boyer and Moore's theorem prover along with other heuristics [1, 2] and has been successfully used to prove many nontrivial theorems as well as reason about hardware and software.

Inspired by Boyer and Moore's work, Zhang, Kapur, and Krishnamoorthy [13, 12] introduced a cover set method for designing induction schemes for automating proofs by induction from equations. This method has been implemented in the theorem prover *Rewrite Rule Laboratory (RRL)* [10] and a proof management system *Tecton* built on top of *RRL* [7], and has been used to prove many nontrivial theorems and reason about sequential as well as parallel programs. The cover set method is based on the assumption that a function symbol is defined by using a finite set of terminating (conditional or unconditional) rewrite rules. The termination ordering employed in orienting the rules is used to perform proofs by well-founded induction. The left sides of the rules are used to design different cases of an induction scheme, and recursive calls to the function made in the right side can be used to design appropriate instantiations for generating induction hypotheses. Since the cover set method is based on a well-founded ordering used to define a function possibly specified by using nonconstructors, an induction scheme generated by this method is a generalization of the principle of mathematical induction for natural numbers and the principle of structural induction obtained from the constructors of a data structure.

In this paper, we go a step further. It is the case that many data structures can be represented in many different ways; for example, numbers can be represented by using 0 and s (to stand for the *successor* function) as constructors as well as in terms of 0, 1, and $+$. Some functions, such as gcd, $divides$, and rem, are more convenient and easier to define by using $+$ instead of s. Finite sets can be represented by using the *null* set, a constructor that *inserts* an element, or alternatively, by using the null set, *singleton* sets, and the *union* operation. Similarly, lists can be represented by using the *empty* list and *cons* constructor, or alternatively, the empty list, *singleton* lists, and *append* operator on lists. While reasoning about functions defined on such data structures, there is a need to convert from one representation to the other.

We exhibit the use of semantic information about a data structure for reconciling the use of different representations while attempting a proof of a conjecture. In this paper, we focus on generating an induction scheme, using semantic information, that enables the application of induction hypotheses that otherwise cannot be applied because of different uses of a function. We illustrate this idea with an example using a decision procedure for Presburger arithmetic (the quantifier-free theory of numbers with the addition operation and relational predicates $>$, $<$, $\neq$, $=$, $\geqslant$, $\leqslant$, also called the *linear arithmetic*) for performing semantic analysis.

First we provide an informal review of the cover set method in mechanizing induction, using a simple example. Then we use another related example to discuss the use of the linear arithmetic procedure for performing semantic analysis for reconciling two different uses of a function in a conjecture.

Consider the following definition of the divisibility predicate on natural numbers; the functions 0, s, $<$, $+$, etc. have the usual meaning.

1. $divides(u, 0) = true$,
2. $divides(u, v) = false\ if\ (v < u) \wedge (v \neq 0)$,
3. $divides(u, u + v) = divides(u, v)$.

The above equations can be oriented left to right, and the resulting rewrite rules are terminating.[1] The above defines $divides$ whenever the first argument is non-zero or both the arguments are 0.

A possible conjecture to prove is

$$(P_1):\ divides(x, y + y)\ if\ divides(x, y) \wedge x > 0.^2$$

The reader can verify that if this conjecture is attempted using the principle of mathematical induction, the proof gets quite complicated. However, with the cover set method, from the definition of $divides(x, y)$ two base cases corresponding to the first two rewrite rules are generated, and the induction step is generated by using the third rewrite rule. The cover set based on the definition of $divides$ is

$$(C_1):\ \{\langle\langle u, 0\rangle, \{\}, \{\}\rangle, \langle\langle u, v\rangle, \{\}, \{v < u, v \neq 0\}\rangle,\\ \langle\langle u, u + v\rangle, \{\langle u, v\rangle\}, \{\}\rangle\}.$$

For every rule in the definition, there is a triple in the cover set, whose first component is the tuple corresponding to the arguments of $divides$ in the left side, the second component is a finite set consisting of the arguments of all the recursive calls to $divides$, if any, in the right side of the rule, and the third component is the finite set of conditions from the literals in the condition, if any, of the rule. A triple could have empty second and third components.

The induction scheme generated from $divides(x, y)$ at position 2.2.1[3] for the above conjecture based on the cover set of $divides$ is

$$(I_1):\ \{\langle\langle\{x \rightarrow u, y \rightarrow 0\}, \{\}, \{2.2.1 \leftarrow divides(u, 0)\}\rangle, \{\}\rangle,\\ \langle\langle\{x \rightarrow u, y \rightarrow v\}, \{v < u, v \neq 0\}, \{2.2.1 \leftarrow divides(u, v)\}\rangle, \{\}\rangle,\\ \langle\langle\{x \rightarrow u, y \rightarrow u + v\}, \{\}, \{2.2.1 \leftarrow divides(u, u + v)\}\rangle,\\ \{\langle\{x \rightarrow u, y \rightarrow v\}, \{\}, \{2.2.1 \leftarrow divides(u, v)\}\rangle\}\rangle\}.$$

An induction scheme is a finite set of tuples, each tuple corresponding to an induction case or subgoal.

The first component of the tuple is used to generate the conclusion in the induction subgoal; the second component is used to generate the induction hypotheses, if any. The first component of the induction case is a 3-tuple, whose first component is a substitution to be made on the conjecture, the second component is a finite set of the conditions for which the induction case is applicable, and the third component is how the subterm being used for generating the induction scheme must be replaced so that the rule of the definition from which the induction case is generated, can be applied. The second component of the induction case is a finite set of 3-tuples for generating the induction hypotheses in the same way as the conclusion is generated. An induction case in which the second component is the empty set corresponds to a base case.

Note that the cover set C_1 covers all the values of *divides* where the first argument is non-zero or both the arguments are zero. Hence the induction scheme generated by the cover set can be used to prove properties under the condition that the first argument of *divides* is non-zero.

For the above conjecture, the first base case is obtained by using the first tuple in the induction scheme (it comes from the first rule), and the substitutions for the variables are $x = u$, $y = 0$:

$$divides(u, 0 + 0) \text{ } if \text{ } divides(u, 0) \wedge u > 0.$$

This formula simplifies to true using the definition of $+$ and the first rule. The second base case is obtained from the second tuple (and it comes from the second rule), and the substitutions for the variables are $x = u$, $y = v$ under the condition $v < u$ and $v \neq 0$:

$$divides(u, v + v) \text{ } if \text{ } divides(u, v) \wedge (u > 0) \wedge (v < u) \wedge v \neq 0,$$

which trivially simplifies to true, since $divides(u, v)$ in the condition simplifies to *false* under the condition $v < u$ and $v \neq 0$ using the second rule.

The induction step comes from the third tuple in the induction scheme (and it comes from the third rule); the substitution for the variables in the conclusion are $x = u$, $y = u + v$:

$$divides(u, (u + v) + (u + v)) \text{ } if \text{ } divides(u, u + v) \wedge u > 0,$$

with the substitutions for the variables in the induction hypothesis coming from the second component as $x = u$, $y = v$:

$$divides(u, v + v) \text{ } if \text{ } divides(u, v) \wedge u > 0.$$

Using the associativity and commutativity properties of $+$ and applying the third rule thrice, we can reduce the conclusion above to

$$divides(u, v + v) \text{ } if \text{ } divides(u, v) \wedge u > 0,$$

which is the induction hypothesis. So the conjecture is proved.

The reader would have noticed how using the well-founded ordering suggested by the definition of *divides* led to an induction hypothesis that turned out to be useful in proving this conjecture.

Now consider a related conjecture:

$$(P_2)\colon\ divides(2,x) = not(divides(2,s(x))),$$

where 2 is an abbreviation for $s(s(0))$.[4] Using the cover set of *divides*, we generate the following induction scheme for $divides(2,x)$ at position 1:

$$\begin{aligned}(I_2)\colon\ &\{\langle\langle\{x \to 0\},\{\},\{1 \leftarrow divides(2,0)\}\rangle,\{\}\rangle,\\ &\langle\langle\{x \to v\},\{v < 2, v \neq 0\},\{1 \leftarrow divides(2,v)\}\rangle,\{\}\rangle,\\ &\langle\langle\{x \to 2+v\},\{\},\{1 \leftarrow divides(2,2+v)\}\rangle,\{\langle\{x \to v\},\{\},\\ &\quad\{1 \leftarrow divides(2,v)\}\rangle\}\rangle\}.\end{aligned}$$

There are two base cases: $divides(2.0) = not(divides(2,s(0)))$, which is proved by using Rules 1 and 2; and similarly, $divides(2,v) = not(divides(2,s(v)))$ *if* $v < 2 \wedge v \neq 0$, which simplifies to $not(divides(2,s(1))) = false$, since $v \neq 0 \wedge v < 2$ implies $v = 1 = s(0)$.

If we check for applicability of definitions and lemmas modulo the theory of linear arithmetic, rewriting can get very expensive. Since rewriting is a primitive operation in a rewrite-based theorem prover such as *RRL*, performing rewriting modulo a theory such as linear arithmetic can slow the theorem prover considerably. It is thus necessary to use the linear arithmetic procedure in a judicious manner to widen the scope of the cover set method and, at the same time, maintain efficiency. In this paper, we do not rewrite modulo the linear arithmetic theory.

We discuss how the subgoal $divides(2,s(1))$ can be proved without rewriting modulo linear arithmetic after considering the induction step generated from the third case in the induction scheme.

For the induction step, the conclusion is

$$divides(2,2+v) = not(divides(2,s(2+v))),$$

if one assumes the induction hypothesis: $divides(2,v) = not(divides(2,s(v)))$.

Rule 3 is now applicable on the left side of the conclusion. But what about the right side? Using semantic information about natural numbers, s and $+$, one could see that $s(2+v) = 2+s(v)$[5], so Rule 3 is applicable to $not(divides(2,s(2+v)))$ to give $not(divides(2,s(v)))$. As stated above, for reasons of efficiency, we do not wish to rewrite modulo the theory of linear arithmetic, so we achieve this by *merging* induction schemes for the two different occurrences of *divides* in the conjecture. So an induction scheme for $divides(2,s(x))$ at position 2.1 is also generated using the cover set of *divides*. The induction schemes for $divides(2, x)$ and $divides(2, s(x))$ are then merged to give the scheme

$$\begin{aligned}(I_3)\colon \{&\langle\langle\{x \to 0\}, \{\}, \{1 \leftarrow divides(2,0), 2.1 \leftarrow divides(2, s(0))\}\rangle, \{\}\rangle,\\ &\langle\langle\{x \to 1\}, \{\}, \{1 \leftarrow divides(2,1), 2.1 \leftarrow divides(2, 2+0)\}\rangle, \{\}\rangle,\\ &\langle\langle\{x \to 2+v\}, \{\}, \{1 \leftarrow divides(2, 2+v),\\ &\qquad\qquad 2.1 \leftarrow divides(2, 2+(v+1))\}\rangle,\\ &\quad \{\langle\{x \to v\}, \{\}, \{1 \leftarrow divides(2, v),\\ &\qquad\qquad 2.1 \leftarrow divides(2, v+1)\}\rangle\}\rangle\}.\end{aligned}$$

The first base case is proved as it was done before. The second base case becomes $divides(2, 1) = not(divides(2, 2+0))$, which is established now by Rules 2 and 3 of the definition of *divides*.

In the induction step, the conclusion is

$$divides(2, 2+v) = not(divides(2, 2+(v+1))),$$

with the hypothesis $divides(2, v) = not(divides(2, v+1))$. The conclusion, by the use of Rule 3, is reduced to

$$divides(2, v) = not(divides(2, v+1)),$$

to which now the hypothesis is applicable. The reason for keeping replacements for subterms being used for generating induction schemes should be evident. In the above conclusion, the subterm at position 1 is already in the form $divides(2, 2+v)$ so that Rule 3 in the definition of *divides* is directly applicable; a similar remark applies for the subterm $divides(2, 2+(v+1))$ at position 2.1.

As the reader must have observed, such semantic analysis can be performed in the case of natural numbers by using the linear arithmetic decision procedure, since reasoning about 0, s, $+$, $=$, $\geqslant$ needs to be performed.

In this paper, we show how Presburger arithmetic can be used in mechanizing induction to reconcile different representations of natural numbers – the one using successor and the other using $+$, $<$. We discuss how a decision procedure for Presburger arithmetic, also called LA, allows one to go back and forth between these two representations in a judicious manner for generating appropriate induction hypotheses. In this way, more and more aspects of proofs by induction can be automated. We discuss how LA can be used to elaborate on a cover set of a function to generate appropriate instantiations for variables in a conjecture to automate its proof.

We also illustrate how semantic analysis using LA is helpful in checking completeness of functions defined using $+$ and $<$ and completeness of a cover set, as well as for merging induction schemes and generalization heuristics in the context of mechanizing proofs by induction.

The main reason for focusing on a linear arithmetic procedure is that it is already integrated in several theorem proving systems such as *Nqthm*, *PVS*, *NEVER*, and *RRL*. Currently in *RRL*, the linear arithmetic procedure is used to prove simple facts or discharge conditions arising in lemmas or definitions [9]. In this paper, we are proposing another important use of the linear arithmetic procedure. The approach presented in this paper should apply to other data structures with multiple representations such as finite lists, finite sequences, finite sets, and multi-sets and for which it is possible to devise decision procedure (or heuristics) to go back and forth among different representations.

2. Cover, Sets, Induction Schemes, Linear Arithmetic

In this section we show how a linear arithmetic decision procedure LA can be used to generate induction schemes. We first give definitions of a cover set and an induction scheme based on a cover set; then we illustrate how on certain classes of conjectures, the cover set method fails because it relies solely on syntactic unification (i.e., unification modulo the empty theory). Linear arithmetic can be used to remedy the problem, since LA can be used to perform semantic unification. We motivate the algorithm using the running example of *divides* given in the introduction. An algorithm for generating an induction scheme for a conjecture from a cover set is given.

2.1. Cover Sets

Given a set F of function symbols and X of variables, $T(F, X)$ denotes the set of terms constructed by using these. Given a term t, by $Vars(t)$, we denote the set of variables that occur in t. A term t whose outermost function symbol is f is called an f-term. An f-term of the form $f(x_1, \ldots, x_n)$ where x_i's are distinct variables is called a *basic* f-term. A substitution σ is a finite map from variables to terms written as $\{x_1 \to s_1, \ldots, x_k \to s_k\}$, $x_i \in X$ and $s_i \in T(F, X)$, $1 \leqslant i \leqslant k$. $\sigma \circ \tau$ denotes the composition of substitutions σ and τ. Given a substitution σ, E_σ denotes σ viewed as equations: $\sigma = \{x_1 \to s_1, \ldots, x_k \to s_k\}$, $E_\sigma = \{x_1 = s_1, \ldots, x_k = s_k\}$. $=_E$ is used to denote equality among terms with respect to a set of (possibly conditional) equations E.

A definition of a function symbol $f \in F$ is a finite set of conditional equations of the form $l_i = r_i$ if $cond_i$, $1 \leqslant i \leqslant m$, where $cond_i$ is optional; l_i is an f-term from $T(F, X)$, r_i is a term in $T(F, X)$, and $cond_i$ is a conjunction of literals built out of terms in $T(F, X)$ such that $Vars(r_i) \subseteq Vars(l_i)$ and $Vars(cond_i) \subseteq Vars(l_i)$. Further it is assumed that each conditional equation in a definition can be oriented into a terminating rewrite rule, i.e., there exists a well-founded reduction ordering $\succ$ over $T(F, X)$ such that $l_i \succ r_i$ and $l_i \succ cond_i$ [4].

Constructors of a data structure need not be free (as in the case of integers and finite sets). It is assumed that constructor relations can be expressed as a finite set of equations. A function may be defined by using nonconstructor symbols.

Let E be a finite set of equations possibly relating constructors and consisting of definitions of nonconstructors used in defining other nonconstructors.

A cover set C_f for a function f with a definition D is a finite set of triples where each triple is derived from a conditional rewrite rule in D. For a conditional rule $l \rightarrow r$ *if cond*, the first component of the triple is the tuple of arguments of f in l, the second component is a set consisting of all tuples serving as arguments to f in r, and the third component is the set of all literals in $cond$. A triple could have its second and third components be the empty set. $Vars(C_f)$ denotes the variables of a cover set C_f.

A cover set C_f for a function $f(x_1, \ldots, x_n)$ is *complete* if and only if for any n-tuple $\langle e_1, \ldots, e_n \rangle$ of ground constructor terms over domains of f, there exists a cover set triple, $c_i = \langle \langle s_1, \ldots, s_n \rangle, \{\ldots, \langle s'_1, \ldots, s'_n \rangle, \ldots\}, cond \rangle$ and a ground substitution σ such that $\sigma(s_i) =_E e_i$, $1 \leqslant i \leqslant n$, and $\sigma(cond) =_E true$. The triple c_i is said to *cover* the tuple $\langle e_1, \ldots, e_n \rangle$. If a cover set C_f of a function is complete, then it follows that f is completely defined.

If a function definition is given solely using constructors, i.e., all the rules in the definition involve only the function symbol being defined and constructor symbols, the definition is called *constructor based*. A cover set generated from a constructor-based definition contains terms only over constructors. Further, if there are no relations on constructors, implying E is the empty set, then in the definition of a complete cover set, $=_E$ is replaced by $=$, and matching can be used. Contextual rewriting with linear arithmetic procedure implemented in [10, 9] is used to discharge the conditions while checking completeness of cover sets. Henceforth, unless stated explicitly, all the cover sets of the function symbols are assumed to be complete.

2.2. Induction Schemes

Given a conjecture C, different nonvariable subterms in C may suggest different ways of performing induction on variables in C. *RRL* currently supports a variety of heuristics to choose an appropriate induction when there are multiple alternatives. Each of these possible alternatives can be specified as an induction scheme. We first discuss the induction scheme generated from a basic f-term $f(x_1, \ldots, x_n)$ appearing in C. Later we discuss how the induction scheme can be generated from a nonbasic f-term $f(t_1, \ldots, t_n)$ of C, where t_i's are not necessarily variables.

The induction scheme suggested by a basic f-term, $f(x_1, \ldots, x_n)$ in C is derived directly from a cover set C_f of f, since it is similar to C_f. Such an induction scheme is called *basic*. For example, the induction scheme I_1 in the Introduction is obtained from $divides(x, y)$ and its cover set, and is basic. As discussed earlier, an induction scheme is a finite set of induction cases of the form

$$\langle \langle \sigma_c, cond_c, repl_c \rangle, \{\ldots, \langle \theta_i, cond_i, repl_i \rangle, \ldots\} \rangle,$$

where $\langle \sigma_c, cond_c, repl_c \rangle$ is used to generate the induction conclusion and each triple in $\{\ldots, \langle \theta_i, cond_i, repl_i \rangle, \ldots\}$ is used to generate an induction hypothesis. Each of these components of an induction case is obtained from a cover set triple $\langle \langle s_1, \ldots, s_n \rangle, \{\ldots, \langle s_1^i, \ldots, s_n^i \rangle, \ldots\}, cond \rangle$ in a cover set C_f of f as follows: $\sigma_c = \{x_1 \to s_1, \ldots, x_n \to s_n\}$, $cond_c = cond$, and $repl_c = \{p \leftarrow f(s_1, \ldots, s_n)\}$, where p is the position of $f(x_1, \ldots, x_n)$ in C; similarly, $\theta_i = \{x_1 \to s_1^i, \ldots, x_n \to s_n^i\}$, $cond_i = cond$, and $repl_i = \{p \leftarrow f(s_1^i, \ldots, s_n^i)\}$.

The substitutions σ_c and each of the substitutions θ_i are linked through the variables shared among the left-hand side and the recursive calls on the right-hand side of the rule from which the cover set triple is derived. The variables whose substitutions are not invariant across the induction conclusion and the hypotheses are called *induction variables*. In the induction scheme I_1 generated from the cover set of $divides(x, y)$, x, y are both induction variables. Given an induction scheme ϕ, let $indvars(\phi)$ denote the induction variables of ϕ and $Vars(\phi)$ denote all the variables substituted for in ϕ.

The induction subgoal corresponding to the above induction case is[6]

$$(\sigma_c(C[repl_c]) \text{ if } cond_c) \text{ if } \left\{ \bigwedge_i \theta_i(C[repl_i]) \text{ if } cond_i \right\}.$$

2.2.1. *Completeness and Soundness of Basic Induction Schemes*

An induction scheme can be used for an inductive proof attempt of a given conjecture provided it is *complete* and *sound*.[7] The completeness and soundness of a basic induction scheme is directly linked to those of the underlying cover set from which it was obtained. A complete cover set results in a complete induction scheme. For constructor-based definitions with free constructors, the soundness of a basic induction scheme follows from the reduction ordering $\succ$ used to prove the termination of the definition of f.

When there are relations among constructors or the definition of f uses non-constructors, i.e., E above is non-empty, a reduction ordering $\succ$ preserving the congruence relation $=_E$ must be used for ensuring that the arguments appearing in the recursive calls on the right side (and/or condition) of each rule in the definition of f is lower than its left side. A reduction ordering $\succ$ preserves $=_E$ iff $u =_E v \succ s =_E t$ implies $u \succ t$ for all terms u, v, s, t. We use $\succ_E$ to denote such an ordering. For any two terms t_1 and $t_2, t_1 \succ_E t_2$ iff for any ground substitution σ and $t =_E \sigma(t_1)$, $t \succ \sigma(t_2)$ [13, 12].

Alternatively, only those instantiations of every rule defining f can be used for generating the cover set for which the recursive calls to f in the right side and the condition are lower than the left side with respect to $\succ_E$. In order to ensure this, the third component of a cover set triple *cond* should be strengthened to include an additional *termination condition* ensuring that every n-tuple in the second component in the triple is lower than the n-tuple appearing in the first component. This approach would be adopted later in the section when a linear arithmetic decision procedure is used for generating cover sets and induction schemes.

THEOREM 1. *Induction scheme* ϕ *generated from a basic* f*-term* $t = f(x_1, \ldots, x_n)$ *of a conjecture* $C(\ldots, t, \ldots)$ *and a complete cover set* C_f *is complete and sound.*

Proof. Completeness: Since C_f is complete, any n-tuple $\langle e_1, \ldots, e_n \rangle$ of ground constructor terms is *covered* by some cover set triple

$$c_i = \langle \langle s_1, \ldots, s_n \rangle, \{\ldots, \langle s_1^j, \ldots, s_n^i \rangle\}, cond \rangle,$$

i.e., for a ground substitution β,

$$\beta(f(s_1, \ldots, s_n)) =_E f(e_1, \ldots, e_n) \quad \text{and} \quad \beta(cond) =_E true,$$

and it is therefore covered by the corresponding induction case

$$\phi_i = \langle \langle \sigma_c, cond_c, repl_c \rangle, \{\ldots, \langle \theta_j, cond_j, repl_j \rangle, \ldots\} \rangle,$$

where $\sigma_c = \{x_1 \to s_1, \ldots, x_n \to s_n\}$, $\theta_j = \{x_1 \to s_1^j, \ldots, x_n \to s_n^j\}$, $cond_c = cond_j = cond$.

Soundness: By contradiction. Assume that $C(\ldots, t, \ldots)$ is not true but the subgoal obtained from each induction case of ϕ is true. Let $\langle e_1, \ldots, e_n \rangle$ be the smallest n-tuple of ground constructor terms with respect to the ordering $\succ_E$ used to prove the termination of f, that is a counterexample to $C(\ldots, t, \ldots)$, i.e., $\beta \circ \sigma_c(C) =_E false$, where $\beta(f(s_1, \ldots, s_n)) =_E f(e_1, \ldots, e_n)$. The induction subgoal corresponding to ϕ_i given above, $(\sigma_c(C)$ *if* $cond)$ assuming $(\bigwedge_j (\theta_j(C)$ *if* $cond)$ is *true*. Since $\beta \circ \sigma_c(C)$ is equivalent to *false* and $\beta(cond) =_E true$, $\beta \circ \theta_j(C)$ is *false* for some j, implying that there is a smaller counterexample to C, since $\beta(f(s_1^j, \ldots, s_n^j))$ is lower in the ordering $\succ_E$ than $f(e_1, \ldots, e_n)$ (by the definition of a cover set, $f(s_1, \ldots, s_n) \succ_E f(s_1^j, \ldots, s_n^j)$ for any j, and since $\succ_E$ is preserved under $=_E$, $f(e_1, \ldots, e_n) \succ_E f(\beta(s_1^j), \ldots, \beta(s_n^j)))$. And, this is a contradiction. □

2.2.2. *Nonbasic Induction Schemes*

The completeness of an induction scheme ϕ generated from a nonbasic f-term $f(t_1, \ldots, t_n)$ ensures that every ground instance of t_i's is *covered* by some induction case of ϕ.

DEFINITION. *Complete induction schemes*: An induction scheme ϕ generated from an f-term $f(t_1, \ldots, t_n)$ is *complete* iff for every ground substitution β there is an n-tuple $\langle e_1, \ldots, e_n \rangle$ of ground constructor terms over the domains of f such that $\beta(f(t_1, \ldots, t_n)) =_E f(e_1, \ldots, e_n)$, there exist an induction case ϕ_j: $\langle\langle\sigma_c, cond_c, repl_c\rangle, \{\ldots, \langle\theta_i, cond_i, repl_i\rangle, \ldots\}\rangle$ and a ground substitution γ such that $\gamma \circ \sigma_c(f(t_1, \ldots, t_n)) =_E f(e_1, \ldots, e_n)$, and $\gamma(cond_c) = true$.

Soundness of an induction scheme ϕ guarantees that if a conjecture C was proved by induction using ϕ, then C is indeed a theorem. More precisely, an induction scheme ϕ generated from an f-term $f(t_1, \ldots, t_n)$ of a given conjecture C is sound iff a proof of C by induction on ϕ implies that for every n-tuple $\langle e_1, \ldots, e_n\rangle$ of ground constructors such that $\beta(f(t_1, \ldots, t_n)) =_E f(e_1, \ldots, e_n)$ for a ground substitution β, and $\beta(C) = true$.

Below we give an algorithm for generating an induction scheme from a non-basic f-term $t = f(t_1, \ldots, t_n)$ appearing in a conjecture C from a cover set C_f of f. The algorithm uses unification to generate different cases of the induction scheme from C_f. If $t = f(x_1, \ldots, x_n)$, it generates a basic induction scheme discussed above. In order to keep the presentation simple, the algorithm is given assuming the definition of f to be constructor based and that there are no relations on constructors (E is empty). The algorithm can be generalized to consider an arbitrary E if syntactic unification (unification modulo the empty theory) below is replaced by $=_E$ unification. Of course, $=_E$ unification must be decidable and finitary, and each of the most general unifiers (*mgus*) in Steps 2(a) is used to generate an induction case, and every *mgu* in Step 2(b) is used to generate an induction hypothesis.

2.2.3. *Algorithm for Generating an Induction Scheme*

- **Input**: A conjecture C of the form $l = r\ if\ cond$, a definition D of an n-ary function f and a term $t = f(t_1, \ldots, t_n)$ at a position p in C.
- **Output**: An induction scheme based on D.
- **Method**:
  1. *Initialize*: Compute the cover set C_f from D.
  2. *Compute the induction scheme*: For each cover set triple, $c = \langle\langle s_1, \ldots, s_n\rangle, \{\ldots, \langle s_1^i, \ldots, s_n^i\rangle, \ldots\}, cond\rangle$ in C_f do:
     a) *Generate induction conclusion*: Let σ be the *mgu* of t and $s = f(s_1, \ldots, s_n)$. The induction conclusion is $\langle\sigma_c, cond_c, repl_c\rangle$ where σ_c is the restriction of σ on $Vars(t)$; $cond_c = \sigma(cond)$; $repl_c = \{p \leftarrow \sigma(s)\}$.
     b) *Generate the induction hypotheses*: Let σ_i be the *mgu* of t and $s' = f(s_1^i, \ldots, s_n^i)$ for some i. The ith induction hypothesis based on σ_i is: $\langle\theta_i, cond_i, repl_i\rangle$ where θ_i is the restriction of σ_i on $Vars(t)$; $cond_i = \sigma_i(cond)$; $repl_i = \{p \leftarrow \sigma_i(s')\}$.

In the above algorithm for generating an induction scheme, if unification of t with $s = f(s_1, \ldots, s_k)$ (or $f(s_1', \ldots, s_k')$) fails, then the safest conclusion is that the cover set method has failed to derive an induction scheme corresponding to the term under consideration (especially if t has occurrences of nonconstructors). If t_i's are constructor terms, then the above algorithm based on syntactic unification can be used to generate a sound and complete scheme from an f-term t as illustrated by the following theorem.

THEOREM 2. *The induction scheme ϕ generated from $t = f(t_1, \ldots, t_n)$ of a conjecture $C(\ldots, t, \ldots)$ and a constructor-based complete cover set C_f, where t_i, $1 \leqslant i \leqslant n$, is a constructor term, assuming no relations on constructors, is complete and sound.*

Proof. Completeness: Without loss of generality assume that

$$Vars(t) \cap Vars(C_f) = \{\}.$$

Since the cover set C_f is complete, for any n-tuple $\langle e_1, \ldots, e_n \rangle$ of ground constructor terms, there exist a ground substitution β and a cover set triple

$$c_i = \langle \langle s_1, \ldots, s_n \rangle, \{\ldots, \langle s_1^k, \ldots, s_m^k \rangle, \ldots\}, cond \rangle$$

such that $\beta(s) = e$ and $\beta(cond) = true$ where $s = f(s_1, \ldots, s_n)$, and $e = f(e_1, \ldots, e_n)$. Let ϕ_i be the induction case corresponding to c_i with the conclusion $\langle \delta_c, \delta(cond), \{p \leftarrow f(\delta(s_1), \ldots, \delta(s_n))\} \rangle$ where δ is the *mgu* of s and t, δ_c is the restriction of δ on $Vars(t)$ and p is the position of t in C.

If for some ground substitution γ, $\gamma(t) = e$, then s and t are unifiable, and $\beta' = \beta \cup \gamma$ is a unifier of s and t ($\beta'(s) = e = \beta'(t)$ since $Vars(s) \cap Vars(t) = \{\}$). So, $\beta' = \theta \circ \delta$ for some substitution θ. Further, since $\beta(cond) = true$, $\beta'(cond) = true$. This implies that $\theta \circ \delta_c(t) = e$ and $\theta \circ \delta(cond) = true$, and therefore $\langle e_1, \ldots, e_n \rangle$ is *covered* by the conclusion, of the induction case ϕ_i. Hence the induction scheme ϕ is complete.

Soundness: By contradiction along the same lines as that of Theorem 1. □

2.3. LINEAR ARITHMETIC AND COVER SETS

When we fail to syntactically unify an arbitrary f-term t with s, no induction case corresponding to the associated cover set triple is generated. If t and/or s involves nonconstructor terms, however, this does not preclude the existence of an n-tuple $\langle e_1, \ldots, e_n \rangle$ of ground constructor terms over the domains of f, which is equivalent to an instance of t as well as an instance of s. Such an n-tuple would not be *covered* by any of the induction cases of the scheme generated, and hence the resulting scheme would be incomplete and, therefore, unsound. This problem can be avoided if unification of t and s is performed modulo a theory containing function symbols other than f occurring in t as well as in s. Consider, for example, the data structure of numbers (natural numbers or integers); s, t could involve $+$. In the next two sections we describe how a linear arithmetic decision procedure can be used to carry out such unification modulo the theory of linear arithmetic for generating complete and sound induction schemes for f-terms whose arguments are terms over the theory of linear arithmetic, and the function f is defined by using terms over the same theory.

The theory of linear arithmetic denotes the quantifier-free first order theory of numbers (integers or natural numbers), numeric variables, the arithmetic operations (successor(s), predecessor(p) and addition($+$)) with the arithmetic relations ($>, <, \neq, =, \geqslant, \leqslant$). A linear term is either a number or a variable or is of the form $f(t_1, \ldots, t_n)$ where f is an n-ary arithmetic operation and t_i, $1 \leqslant i \leqslant n$, are linear terms. If p is an n-ary arithmetic relation, $p(t_1, \ldots, t_n)$, where t_i's are linear terms denoting a linear atom. Linear literals are either linear atoms or negations of linear atoms. $=_{LA}$ is used to denote the equality amongst terms with respect to the linear arithmetic theory.

Let $CE = \{s_i = t_i \mid 1 \leqslant n\}$ be a set of equations with a context *cond* where s_i's and t_i's are linear terms and *cond* is a conjunction of linear literals. Let $\{x_1, \ldots, x_m\}$ be the variables occurring in CE and *cond*. An m-tuple of $\langle e_1, \ldots, e_m\rangle$ of numbers is a solution to CE under *cond* iff $\sigma(cond) =_{LA} true$ and $\sigma(s_i) =_{LA} \sigma(t_i)$, for all $1 \leqslant i \leqslant n$, where $\sigma = \{x_j - e_j \mid 1 \leqslant j \leqslant m\}$.

By the linear arithmetic procedure LA, we assume the existence of an algorithm specified as follows.

- *Input*: A set of equations CE: $\{s_i = t_i \mid 1 \leqslant i \leqslant n\}$ and a context *cond*, where s_i's and t_i's are linear terms and *cond* is a conjunction of linear literals.
- *Output*: No, if CE cannot be solved. Otherwise, yes, with a solved form of CE as a substitution σ of the form $\{x_j \rightarrow u_j\}$ satisfying the occur-check, where u_j are linear terms possibly involving new variables introduced while solving CE, and feasibility constraints c_c, a conjunction of linear literals restricting values that variables can take.
- *Completeness and Soundness*: The input and the output of the LA are related such that any m-tuple $e = \langle e_1, \ldots, e_m\rangle$ of numbers is a solution to the equations CE with context *cond* if and only if e is a solution to E_σ with context $\{\sigma(cond), c_c\}$.

For equations, a solved form, if any, can be obtained by using an algorithm for solving linear diophantine equations, and conditions on variables, if needed, can be included in feasibility constraints. We assume that given a linear literal of the form $x \leqslant c$ where c is a constant, it is possible to enumerate by using the procedure the various values for x, $x = c_i$, where $c_i \leqslant c$. The requirement of completeness is only an algorithm for solving a set of linear equations under a context. Further, since the input is a restricted subset of universally quantified linear arithmetic theory, it can be shown that a complete procedure would either detect the unsolvability of the given set of equations under the given context or would lead to a solution that is expressible as a single substitution along with feasibility constraints.

DEFINITION. *LA-based function definitions*: A definition D of an n-ary function f is LA-based iff for each rule $l \rightarrow r$ *if cond* of D, the arguments to f in l and r are linear terms; *cond* is a conjunction of linear literals.

Most of the definitions in this paper are LA-based definitions. For example, the definition of *divides* in the introduction is an LA-based definition.

For using LA to do semantic analysis for generating an induction scheme from a cover set, there is a need to modify the definition of cover set slightly. Since we propose to use LA to perform semantic unification of the terms in the conjecture with those in the cover set generated from definitions of functions, it is not enough if the definitions of the functions are oriented into terminating rules based on a well-founded reduction ordering. Such orderings need not be preserved under semantic transformations. For instance, given a rule such as $f(x+0) \rightarrow f(x)$ in the definition of f, semantic unification with a term $f(n)$ in a conjecture using LA can replace the term $x+0$ by x, which is semantically equivalent to it, to produce an induction case of the form $\langle\langle\{n \rightarrow x\}, \{\}, \{\}\rangle, \{\langle\{n \rightarrow x\}, \{\}, \{\}\rangle\}\rangle$, which is unsound. To avoid this, we require that the reduction ordering $\succ$ preserve the semantic congruence relation amongst numbers induced by $=_{LA}$. For soundness, for every rule $l \rightarrow r$ *if cond*, where l is an f-term and the arguments to f are linear terms, arguments in recursive calls of f in r (as well as *cond*) must be lower than the arguments to f in l in a semantic well-founded ordering over numbers. This additional requirement is incorporated in every triple in the cover set of f by strengthening its condition component *cond* to include a *termination* condition.

For natural numbers, a lexicographic order on k-tuples using $>$ works as a well-founded order. For integers, a lexicographic order using $>$ on absolute values of linear terms would work. A semantic well-founded ordering on linear terms must be preserved under substitutions. A well-founded semantic ordering on linear terms, $\succ_{LA}$ is defined as: $t_1 \succ_{LA} t_2$ iff for every substitution σ from variables to numbers, $|\sigma(t_1)| > \sigma(t_2)|$ where $|t|$ denotes the absolute value of a linear term t. $\succ_{lex}$ denotes the lexicographic extension of $\succ_{LA}$ to tuples of linear terms. For example, for the third rule defining *divides*, $\langle u, u+v\rangle > \langle u, v\rangle$, which is the case only if $u + v > v$; so the termination condition $u + v > v$ is added to the third component in the triple corresponding to the third rule in the cover set for *divides*.

DEFINITION. *LA-based cover sets*: A cover set C_f of an n-ary function f is LA-based iff for any cover set triple $c_i = \langle\langle s_1, \ldots, s_n\rangle, \{\ldots, \langle s_1^k, \ldots, s_n^k\rangle, \ldots\},$ $cond \wedge T\rangle$ of C_f, s_i and s_i^k, $1 \leqslant i \leqslant n$, for any k, are linear terms and *cond* is a conjunction of linear literals and T, the termination condition, is a conjunction of linear literals ensuring that $\langle s_1, \ldots, s_n\rangle \succ_{lex} \langle s_1^k, \ldots, s_n^k\rangle$, for any k.

Based on the above definition, the LA-based cover set of the predicate *divides* introduced in Section 1 gets modified as follows.

$$C_{divides}: \quad \{\langle\langle u, 0\rangle, \{\}, \{\}\rangle, \langle\langle u, v\rangle, \{\}, \{v < u, v \neq 0\}\rangle, \\ \langle\langle u, u+v\rangle, \{\langle u, v\rangle\}, \{u+v > v\}\rangle\}.$$

DEFINITION. *Complete LA-based cover sets*: An LA-based cover set C_f for an n-ary function f is *complete* iff for any n-tuple of numbers $\langle e_1, \ldots, e_n\rangle$ there exist a cover set triple $\langle\langle s_1, \ldots, s_n\rangle, \{\ldots, \langle s'_1, \ldots, s'_n\rangle, \ldots\}, cond \wedge T\rangle$ and a ground substitution σ such that $\sigma(s_i) =_{LA} e_i$ is for all i, $1 \leqslant i \leqslant n$, and $\sigma(cond \wedge T) = true$.

In discussions below, we assume that the termination condition T is integrated into $cond$, the condition governing the cover set triples, and hence it will not be explicitly specified.

2.4. Linear Arithmetic Based Induction Schemes

For generating an induction scheme from an f-term t in a conjecture C and the cover set of f, if the arguments of f in t corresponding to the induction variables are non-variable terms, then the algorithm in Section 2.2.3 often fails. For example, take t to be $divides(x, y+y)$ in the conjecture discussed earlier. It is not possible to derive an induction scheme by using the cover set C_1 because $y+y$ in C cannot be unified with 0 for the first cover set triple itself.

Another possible situation to consider is when the arguments to the term t in the argument positions corresponding to the induction variables are not distinct variables. For example, take t to be $divides(x, x)$ in some conjecture. Again the cover set method fails to generate an induction scheme for this case because x needs to be unified with both u and $u+v$ to generate an induction case from the last triple of the cover set, which is not possible.

Failure in both cases is because of purely syntactic view taken in attempting to unify terms expressed by using 0 and $+$. In the first case, $y+y$ can be unified with 0 by using semantic analysis, giving the most general unifier $\{y \rightarrow 0\}$. Similarly, in the second case also, x can be unified with both u and $u+v$, giving the most general unifier $\{x \rightarrow u, v \rightarrow 0\}$. There is thus a need to relate arbitrary terms expressed by using 0, s, $+$, and this can be achieved by using the linear arithmetic procedure (LA). With LA, both the cases above can be solved.

Using LA, we can generalize the algorithm in Subsection 2.2.3, which uses syntactic unification for generating the induction scheme for a conjecture C, to perform semantic unification of a selected subterm $t = f(t_1, \ldots, t_n)$ and the cover set associated with the outermost symbol f of t. Given a cover set triple, $c = \langle\langle s_1, \ldots, s_n\rangle, \{\ldots, \langle s^j_1, \ldots, s^j_n\rangle, \ldots\}, cond\rangle$, constraint equations $\{t_i = s_i \mid 1 \leqslant i \leqslant n\}$ are set up; LA is invoked to produce substitution σ of $Vars(t) \cup Vars(c)$ with context $cond$. If the third component of c is empty, then the context is assumed to be $true$. The restriction of σ on $Vars(t)$ produces the substitution corresponding to the induction conclusion; similarly, substitution σ_j of $Vars(t) \cup Vars(c)$ for induction hypotheses are generated by invoking LA on the constraint equations $\{t_i = s_i^j \mid 1 \leqslant i \leqslant n\}$ with context *cond*. Restriction of σ_j on $Vars(t)$ is the substitution corresponding to the jth induction hypothesis.

Since we wish to avoid rewriting modulo linear arithmetic, we associate with the induction conclusion and with each of the induction hypotheses a set of *positional* replacements of the form $\{p \leftarrow \sigma(f(s_1, \ldots, s_k))\}$ and $\{p \leftarrow \sigma_i(f(s'_1, \ldots, s'_k))\}$, respectively, where p is the position of the term t in the conjecture C. These replacements enable the use of the rules from the definition of f for simplification, as seen in the case of conjecture P_2 earlier.

For example, consider the term $divides(2, x)$ at position 1 in the conjecture P_2 presented in the introduction. From the first cover set triple of *divides*, we get $\{x \to 0\}$, as the substitution. From the second triple, the substitution is obtained from the constraints $\{2 = u, x = v\}$ with context $\{v < u, v \neq 0\}$, which using LA, simplify to $\{x \to 1, u \to 2, v \to 1\}$. From the third cover set triple, the constraints corresponding to the induction conclusion are $\{2 = u, x = u + v\}$ with context $\{u + v > v\}$, which simplify to $\{x \to 2 + v, u \to 2\}$.[8] In a similar manner, the substitution corresponding to the induction hypothesis is obtained to be $\{x \to v, u \to 2\}$. Hence the induction scheme is given as

$$\begin{aligned}&\{\langle\langle\{x \to 0\}, \{\}, \{1 \leftarrow divides(2,0)\}\rangle, \{\}\rangle, \langle\langle\{x \to 1\}, \{\},\\ &\qquad\qquad \{1 \leftarrow divides(2,1)\}\rangle, \{\}\rangle,\\ &\langle\langle\{x \to 2 + v\}, \{\}, \{1 \leftarrow divides(2, 2 + v)\}\rangle, \{\langle\{x \to v\}, \{\},\\ &\qquad \{1 \leftarrow divides(2, v)\}\rangle\}\rangle\}.\end{aligned}$$

The induction scheme for the term $divides(2, s(x))$ at position 2.1 in the conjecture is computed as follows. No induction case is generated corresponding to the first cover set triple, since 0 and $s(x)$ cannot be unified using LA. From the second cover set triple, the constraints are $\{u' = 2, v' = s(x)\}$ with context $\{v' < u', v' \neq 0\}$, which simplifies by using LA to $\{x \to 0, u' \to 2, v' \to s(0)\}$. From the third cover set triple, the constraints corresponding to the conclusion are $\{u' = 2, u' + v' = s(x)\}$ with context $\{u' + v' > v'\}$, which simplifies to $\{x \to v' + 1, u' \to 2\}$. The constraints for the hypothesis are $\{u' = 2, v' = s(x)\}$ with context $\{u' + v' > v'\}$, which simplifies to $\{x \to v' - 1, u' \to 2\}$ along with the feasibility constraint $\{v' \geqslant 1\}$.

For $divides(2, s(x))$, the induction scheme is specified as

$$\begin{aligned}&\{\langle\langle\{x \to 0\}, \{\}, \{2.1 \leftarrow divides(2, s(0))\}\rangle, \{\}\rangle,\\ &\langle\langle\{x \to v' + 1, \{\}, \{2.1 \leftarrow divides(2, 2 + v')\}\rangle, \{\langle\{x \to v' - 1\}, \{v' \geqslant 1\},\\ &\qquad \{2.1 \leftarrow divides(2, v')\}\rangle\}\rangle\}.\end{aligned}$$

2.4.1. *Modified Algorithm for Generating an Induction Scheme*

- **Input**: A conjecture C of the form $l = r$ *if cond*, and LA-based definition D of an n-ary function f and a term $t = f(t_1, \ldots, t_n)$ at a position p in C.

- **Output**: An induction scheme based on D.
- **Method**:
  1. *Initialize*: Compute the LA-based cover set $\mathcal{C}_f$ using D.
  2. *Compute induction schemes*: For each cover set triple,
     $c = \langle\langle s_1, \ldots, s_n\rangle, \{\ldots, \langle s'_1, \ldots, s'_n\rangle, \ldots\}, cond\rangle$ in $\mathcal{C}_f$ do:
     a) *Generate induction conclusion*: The induction conclusion of the form $\langle \sigma_c, cond_c, repl_c\rangle$ is generated as follows:
        i) Set up constraint equations $CE = \{t_i = s_i \mid 1 \leqslant i \leqslant n\}$.
        ii) Solve CE using LA with context $\{cond\}$. If solvable, let $\sigma = \{x_i \to u_i\}$ be the substitution obtained that satisfies the occur-check, where u_i are linear terms possibly involving new variables and $x_i \in Vars(t) \cup Vars(c)$. Let c_c be the set of feasibility constraints over $Vars(c)$. No solution corresponds to a vacuous induction case.
        iii) Let μ be the restriction of σ on $Vars(c)$. σ_c is the restriction of σ on $Vars(t)$; $cond_c$ is $\mu(c_c) \cup \mu(cond)$[9]; $repl_c$ is $\{p \leftarrow f(\sigma(s_1), \ldots, \sigma(s_k))\}$.
     b) *Generate the set of induction hypotheses*: The ith hypothesis $\langle \theta_i, cond_i, repl_i\rangle$ is generated as follows:
        i) Set up constraint equations $CE = \{t_i = s'_i \mid 1 \leqslant i \leqslant k\}$. Solve CE using LA as before, with context $\{cond_c, CE_\mu\}$. Let c'_c be the new feasibility constraints obtained, and let σ_i be the new substitution obtained.
        ii) Use σ_i and c'_c to compute θ_i, $cond_i$ and $repl_i$ as before. No solution corresponds to the ith hypothesis not being generated.

The induction scheme output by the above algorithm can be applied to the conjecture as discussed earlier. It should be noted that in the above modified algorithm the cases that do not yield solutions can be ignored because the underlying LA procedure is assumed to be sound and complete.

2.4.2. *Examples*

We illustrate the algorithm on some examples involving the greatest common divisor (gcd) function. The properties illustrated below cannot be proved with the cover set induction method without using LA. The methods implemented for induction in *Nqthm* [1] are also unsuccessful in establishing these properties (these properties can be proved by using hints and other tricks, however). The conjectures are based on the following definition of the gcd function.

1. $gcd(0, x) \to x$, 2. $gcd(x, 0) \to x$,
3. $gcd(x + y, y) \to gcd(x, y)$, 4. $gcd(x, x + y) \to gcd(x, y)$.

The first conjecture that we consider is

(P_3): $gcd(m + m, 2) = 2$.

1. *Initialize*: The LA-based cover set C_{gcd} for gcd is
$\{\langle\langle 0,x\rangle,\{\},\{\}\rangle,\langle\langle x,0\rangle,\{\},\{\}\rangle,\langle\langle x+y,y\rangle,\{\langle x,y\rangle\},\{x+y>x\}\rangle,$
$\langle\langle x,x+y\rangle,\{\langle x,y\rangle\},\{x+y>y\}\rangle\}$.
The conditions governing the last two cover set triples are terminating conditions. For the third rule, $\langle x+y,y\rangle \succ_{lex} \langle x,y\rangle$ iff $x+y>x$. Similarly the last rule defining gcd leads to the termination condition $x+y>y$.
2. Compute the induction scheme:
   a) The first cover set triple would generate the induction case $\langle\langle\{m\rightarrow 0\},\{\},\{1\leftarrow gcd(0,2)\}\rangle,\{\}\rangle$. The second cover set triple would not generate an induction case, since $2=0$ cannot be solved.
   b) With the third cover set triple $\langle\langle x+y,y\rangle,\{\langle x,y\rangle\},\{x+y>x\}\rangle$ to generate the substitution for the induction conclusion, the constraint equations are $\{m+m=x+y, y=2\}$ with context $\{x+y>x\}$, and the substitution obtained is $\{m\rightarrow z+1, x\rightarrow z+z, y\rightarrow 2\}$ where z is a new variable introduced while solving for $m+m=x+2$. The constraint equations for the hypothesis are $\{m+m=x, y=2\}$ with context $\{x=z+z, y=2\}$, and the substitution obtained is $\{m\rightarrow z, y\rightarrow 2\}$. The induction case generated is $\langle\langle\{m\rightarrow z+1\},\{\},\{1\leftarrow gcd(z+z+2,2)\}\rangle,$ $\{\langle\{m\rightarrow z\},\{\},\{1\leftarrow gcd(z+z,2)\}\rangle\}\rangle$.
   c) For the fourth cover set triple, the constraint equations for the conclusion are $\{x=m+m, x+y=2\}$ with context $\{x+y>y\}$. The substitution obtained is $\{m\rightarrow 1, x\rightarrow 2, y\rightarrow 0\}$. The constraint equations for the hypothesis are $\{x=m+m, y=2\}$ with context $\{y=0, x=2\}$, which are unsatisfiable. So there is no induction hypothesis for this case, and the induction case generated is $\langle\langle\{m\rightarrow 1\},\{\},\{1\leftarrow gcd(2,2+0)\}\rangle,\{\}\rangle$.

The scheme generated by the algorithm is

$$\begin{aligned}\{&\langle\langle\{m\rightarrow 0\},\{\},\{1\leftarrow gcd(0,2)\}\rangle,\{\}\rangle,\\ &\langle\langle\{m\rightarrow 1\},\{\},\{1\leftarrow gcd(2,2+0)\}\rangle,\{\}\rangle,\\ &\langle\langle\{m\rightarrow z+1\},\{\},\{1\leftarrow gcd(z+z+2,2)\}\rangle,\\ &\qquad\{\langle\{m\rightarrow z\},\{\},\{1\leftarrow gcd(z+z,2)\}\rangle\}\rangle\}.\end{aligned}$$

Since any pair of numbers $\langle e_1,e_2\rangle$ *covered* by the induction Case 2 is also *covered* by Case 3, Case 2 can be dropped. Using this induction scheme, the conjecture is easily proved.

The next conjecture that we consider is

(P_4): $gcd(m,s(m))=1$.

1. *Initialize*: The LA-based cover set C_{gcd} is the same as before.
2. Compute the induction scheme:
   a) For the first cover set triple, the constraint equations $\{m=0, s(m)=x\}$ are trivially solvable. The substitution obtained is $\{m\rightarrow 0, x\rightarrow 1\}$. The induction case generated is $\langle\langle\{m\rightarrow 0\},\{\},\{1\leftarrow gcd(0,1)\}\rangle,\{\}\rangle$.

b) The second cover set triple does not lead to an induction case, since the constraint equations $\{m = x, s(m) = 0\}$ do not have a solution.

c) The third cover set triple does not lead to an induction case either, since the constraint equations for the conclusion $\{m = x + y, s(m) = y\}$ with context $\{x + y > x\}$ do not have a solution.

d) For the fourth cover set triple, the constraint equations for the conclusion are $\{m = x, s(m) = x + y\}$ with context $\{x + y > y\}$. The substitution obtained is $\{m \to x\}$ with feasibility constraints $\{x > 0\}$. The constraint equations for the hypothesis $\{m = x, s(m) = y\}$ with context $\{y = 1, x > 0\}$ do not have a solution and the hypothesis discarded.

The induction scheme generated is given below. The subgoal generated from the first induction case of this scheme reduces to true by rule 1 in the definition of gcd. The subgoal generated from the second case reduces to $gcd(x, 1) = 1$, which can be established by induction (using the scheme generated from the subterm $gcd(x, 1)$ using LA).

$$\{\langle\langle\{m \to 0\}, \{\}, \{1 \leftarrow gcd(0, 1)\}\rangle, \{\}\rangle,$$
$$\langle\langle\{m \to x\}, \{x > 0\}, \{1 \leftarrow gcd(x, x + 1)\}\rangle, \{\}\rangle\}.$$

2.4.3. *Completeness and Soundness of Schemes*

In Section 2.1, we proved the completeness and the soundness of induction schemes generated from basic f-terms and complete cover sets. It was also shown in Section 2.2 that the induction schemes generated by syntactic unification with the cover set triples for a nonbasic f-term $f(t_1, \ldots, t_n)$ need not be complete even though the underlying cover set is complete. For a nonbasic f-term, complete and sound schemes can be obtained if, instead of syntactic unification, unification is performed modulo the equational theory E, comprising definitions and relations among constructors. This is possible only if unification modulo E is decidable and finitary, which is always not the case. We prove below that induction schemes generated from complete LA-based cover sets by using LA, and an f-term $f(t_1, \ldots, t_n)$, where t_i's are linear terms, by the algorithm in Section 2.4.1, are complete and sound. The proof is along the lines of Theorem 2.

THEOREM 3. *Induction scheme ϕ generated from $t = f(t_1, \ldots, t_n)$ of a conjecture $C(\ldots, t, \ldots)$ and a complete LA-based cover set C_f is complete and sound.*

Proof. Completeness: Without loss of generality assume that

$$Vars(t) \cap Vars(C_f) = \{\}.$$

Since C_f is a complete LA-based cover set, for any n-tuple $\langle e_1, \ldots, e_n\rangle$ of numbers, there exist a ground substitution β and a cover set triple $c_i = \langle\langle s_1, \ldots, s_n\rangle, \{\ldots, \langle s_1^k, \ldots, s_n^k\rangle, \ldots\}, cond\rangle$ such that $\beta(s) =_{LA} e$ and $\beta(cond) = true$, where

$s = f(s_1, \ldots, s_n)$ and $e = f(e_1, \ldots, e_n)$. Let ϕ_i be the induction case corresponding to c_i with the conclusion

$$\langle\langle \delta_c, \delta(cond) \wedge c_c, \{p \leftarrow f(\delta(s_1), \ldots, \delta(s_n))\}\rangle.$$

Let δ be the *mgu* of s and t obtained by solving $CE = \{s_i = t_i \mid 1 \leqslant i \leqslant n\}$ under the context $cond$, using LA, with the feasibility constraints c_c. Then δ_c is the restriction of δ to $Vars(t)$, and p is the position of t in C.

If for some ground substitution γ, $\gamma(t) =_{LA} e$, then s and t are unifiable modulo $=_{LA}$, and $\beta' = \beta \cup \gamma$ is a unifier of s and t ($\beta'(s) =_{LA} e =_{LA} \beta'(t)$, $Vars(s) \cap Vars(t) = \{\}$), and since $\beta(cond) = true$, $\beta'(cond) = true$. Therefore, by the completeness and soundness of LA, $\beta' = \theta \circ \delta$ where $\theta(c_c) = true$, for some substitution θ. This implies that $\theta \circ \delta_c(t) =_{LA} e$ and $\theta \circ \delta(cond) = true$ and $\theta(c_c) = true$ and, therefore, $\langle e_1, \ldots, e_n\rangle$ is *covered* by the induction case ϕ_i corresponding to c_i. Hence ϕ is complete.

Soundness: By contradiction. Assume that $C(\ldots, t, \ldots)$ is not true but the subgoal obtained from each of the induction cases of ϕ is true. Let $\langle e_1, \ldots, e_n\rangle$ be the smallest n-tuple of numbers, with respect to the semantic ordering $\succ_{LA}$ used to prove the termination of the definition of f, that is a counterexample to $C(\ldots, t, \ldots)$, i.e., $\theta \circ \delta_c(C) = false$. The induction subgoal corresponding to ϕ_i given above, $(\delta_c(C)\ if\ (\delta(cond) \wedge c_c))$ assuming $(\bigwedge_k (\theta_k^h(C)\ if\ (\theta_k(cond) \wedge c_c')))$ is $true$. Since $\theta(\delta_c(C))$ is $false$, $\theta(\delta(cond)) = true$ and $\theta(c_c) = true$, $\theta(\theta_k^h(C)) = false$ for some k, implying that there is a smaller counterexample to C, since $f(\theta(\theta_k(s_1^k)), \ldots, \theta(\theta_k(s_n^k)))$ is lower in the ordering $\succ_{LA}$ than c (by definition of LA-based cover set, $s \succ_{LA} f(s_1^k, \ldots, s_n^k))$ for any k and since $\succ_{LA}$ is $=_{LA}$ preserving, $f(\theta(\delta(s_1)), \ldots, \theta(\delta(s_n))) \succ_{LA} f(\theta(\delta(s_1^k)), \ldots, \theta(\delta(s_n^k)))$. A contradiction. □

3. Merging Induction Schemes

Often the induction schemes suggested by the various subterms of a given conjecture share induction variables among each other. An inductive proof attempt of the conjecture based on one of these schemes only is not likely to succeed in such cases. For instance, if $t_1 = f(x, y)$ and $t_2 = g(z, x)$ are any two subterms of a given conjecture where f and g are binary functions defined recursively on both of their arguments, then attempting a proof of the conjecture by induction based only on the scheme suggested by the term t_1 would result in an induction step with the conclusion containing $t_1' = \sigma(f(x, y)) = f(\sigma(x), \sigma(y))$ and $t_2' = \sigma(g(z, x)) = g(z, \sigma(x))$. The choice of induction scheme ensures that the term t_1' can be simplified to match the induction hypothesis, but the same need not be true for the term t_2' because the variable z in t_2' does not get instantiated.

For example, based on the following definition of the *same* predicate, defining equality on *lists* built out of constructors nil and $cons$,

1. $same(nil, nil) \rightarrow true$,
2. $same(nil, cons(u_1, v_1)) \rightarrow false$,
3. $same(cons(u_1, v_1), nil) \rightarrow false$,
4. $same(cons(u_1, v_1), cons(u_2, v_2)) \rightarrow (u_1 = u_2) \wedge same(v_1, v_2)$,

a conjecture that can be attempted is

$$(P_5):\ same(u, w)\ if\ (same(u, v) \wedge same(v, w)).$$

The schemes suggested by the subterms $same(u, v)$, $same(v, w)$, and $same(u, w)$ are three possible candidates for attempting a proof of the conjecture (P_5) by induction. The scheme suggested by the subterm $same(u, v)$ is given as follows.

$$\begin{aligned}(I_4):\ \{&\langle\langle\{u \rightarrow nil, v \rightarrow nil\}, \{\}, \{\}\rangle, \{\}\rangle,\\ &\langle\langle\{u \rightarrow nil, v \rightarrow cons(x_1, y_1)\}, \{\}, \{\}\rangle, \{\}\rangle,\\ &\langle\langle\{u \rightarrow cons(x_1, y_1), v \rightarrow nil\}, \{\}, \{\}\rangle, \{\}\rangle,\\ &\langle\langle\{u \rightarrow cons(x_1, y_1), v \rightarrow cons(x_2, y_2)\}, \{\}, \{\}\rangle,\\ &\quad \{\langle\{u \rightarrow y_1, v \rightarrow y_2, \{\}, \{\}\rangle\}\rangle\}.\end{aligned}$$

The induction scheme suggested by $same(u, w)$ is very similar to the above scheme and can be obtained by replacing the variable v by the variable w in each of the induction cases of the above scheme. The induction scheme suggested by the subterm $same(v, w)$ is given as follows.

$$\begin{aligned}I_5:\ \{&\langle\langle\{v \rightarrow nil, w \rightarrow nil\}, \{\}, \{\}\rangle, \{\}\rangle,\\ &\langle\langle\{v \rightarrow nil, w \rightarrow cons(x'_1, y'_1)\}, \{\}, \{\}\rangle, \{\}\rangle,\\ &\langle\langle\{v \rightarrow cons(x'_1, y'_1), w \rightarrow nil, \{\}, \{\}\rangle, \{\}\rangle,\\ &\langle\langle\{v \rightarrow cons(x'_1, y'_1), w \rightarrow cons(x'_2, y'_2)\}, \{\}, \{\}\rangle,\\ &\quad \{\langle\{v \rightarrow y'_1, w \rightarrow y'_2\}, \{\}, \{\}\rangle\}\rangle\}.\end{aligned}$$

Attempting a proof of (P_5) by induction based on only one of the above schemes, such as (I_4), would result in the induction step case with the conclusion

$$\begin{aligned}&same(cons(x_1, y_1), w)\ if\ same(cons(x_1, y_1),\\ &\quad cons(x_2, y_2)) \wedge same(cons(x_2, y_2), w),\end{aligned}$$

and the hypothesis $same(y_1, w)\ if\ same(y_1, y_2) \wedge same(y_2, w)$. The conclusion simplifies using the last rule in the definition of $same$ to

$$\begin{aligned}&same(cons(x_1, y_1), w)\ if\ (x_1 = x_2) \wedge same(y_1, y_2)\\ &\quad \wedge same(cons(x_2, y_2), w).\end{aligned}$$

Since the conclusion cannot be simplified any further, the hypothesis does not match the conclusion, and the proof attempt by induction fails.[10] The failure is due to the induction variable w not being instantiated in the subterms $same(u, w)$ and $same(v, w)$ of (P_5) in the conclusion, as discussed before. The reader can easily verify that a proof attempt of (P_5) based on the other two schemes would also fail, since one of the variables u or v would remain uninstantiated in the induction step of these proof attempts.

This situation can be remedied if the above two induction schemes (I_4) and (I_5) are merged and an induction scheme that instantiates the induction variables simultaneously in all the terms is generated. Further, merging of schemes also eliminates the need to arbitrarily choose from among competing schemes. We merge the induction scheme (I_4) into the scheme (I_5) by merging each induction case of (I_4) into as many cases of (I_5) as possible based on the following definition of merging of induction cases.

DEFINITION. *Mergeable induction cases*: An induction case

$$i_1 = \langle\langle \sigma_c, cond_c, repl_c\rangle, \{\ldots, \langle\theta_i, cond_i, repl_i\rangle, \ldots\}\rangle$$

merges with an induction case

$$i_2 = \langle\langle \sigma'_c, cond'_c, repl'_c\rangle, \{\ldots, \langle\theta'_j, cond'_j, repl'_j\rangle, \ldots\}\rangle$$

to give an induction case of the form $\langle\langle \sigma, mcond_c, mrepl_c\rangle, MH\rangle$ if the following conditions hold.

- $\sigma_c(x)$ is unifiable with $\sigma'_c(x)$ with the *mgu* δ for all $x \in Vars(\sigma_c) \cap Vars(\sigma'_c)$. $\sigma = (\delta \circ \sigma_c) \cup (\delta \circ \sigma'_c)$; $mcond_c = \delta(cond_c) \cup \delta(cond'_c)$; $mrepl_c = \delta(repl_c) \cup \delta(repl'_c)$.
- The (i, j)th merged hypothesis is:[11] For each triple, $\langle\theta_i, cond_i, repl_i\rangle$ of i_1 the triple $\langle(\delta \circ \theta_i) \cup \gamma'_j \delta(cond_i) \cup \delta(cond'_j), \delta(repl_i)\rangle$ is included in MH if it is consistent. If the second component of i_2 is empty (a basis case), then γ'_j is the restriction of σ'_c on variables appearing only in i_2. Otherwise, γ'_j is the restriction of θ'_j on such variables.
  Similarly, for each triple, $\langle\theta'_j, cond'_j, repl'_j\rangle$ of i_2, the triple $\langle(\delta \circ \theta'_j) \cup \gamma_i, \delta(cond'_j) \cup \delta(cond_i), \delta(repl'_j)\rangle$ is included in MH if it is consistent. If the second component of i_1 is empty, then γ_i is the restriction of σ_c on variables appearing only in i_1. Otherwise, γ_i is the restriction of θ_i on such variables.

In order to merge (I_4) into (I_5), we consider each induction case of (I_4) with all induction cases of (I_5). The first induction case of (I_4) merges only with the first two induction cases of (I_5), since for the other cases the substitutions for the shared variable v, nil and $cons(x, y)$, cannot be unified. The merged cases are

$$\langle\langle\{u \to nil, v \to nil, w \to nil\}, \{\}, \{\}\rangle, \{\}\rangle,$$
$$\langle\langle\{u \to nil, v \to nil, w \to cons(x'_1, y'_1)\}, \{\}, \{\}\rangle, \{\}\rangle.$$

The second induction case of (I_4) merges only with the last two induction cases of (I_5) for the same reasons as above. The merged cases are

$$\langle\langle\{u \to nil, v \to cons(x_2, y_2), w \to nil\}, \{\}, \{\}\rangle, \{\}\rangle,$$
$$\langle\langle\{u \to nil, v \to cons(x_2, y_2), w \to cons(x'_2, y'_2)\}, \{\}, \{\}\rangle,$$
$$\{\langle\{u \to nil, v \to y_2, w \to y'_2\}, \{\}, \{\}\rangle\}\rangle.$$

Similarly for the third induction case of (I_4) the merged cases obtained are

$$\langle\langle\{u \to cons(x_1, y_1), v \to nil, w \to nil\}, \{\}, \{\}\rangle, \{\}\rangle,$$
$$\langle\langle\{u \to cons(x_1, y_1), v \to nil, w \to cons(x_2', y_2')\}, \{\}\{\}\rangle, \{\}\rangle.$$

The final induction case of (I_4) merges with the third and the fourth induction cases of (I_5), and the merged induction cases obtained are

$$\langle\langle\{u \to cons(x_1, y_1), v \to cons(x_2, y_2), w \to nil\}, \{\}, \{\}\rangle,$$
$$\{\langle\langle\{u \to y_1, v \to y_2, w \to nil\}, \{\}, \{\}\rangle\}\rangle,$$
$$\langle\langle\{u \to cons(x_1, y_1), v \to cons(x_2, y_2), w \to cons(x_2', y_2')\}, \{\}, \{\},$$
$$\{\langle\langle\{u \to y_1, v \to y_2, w \to y_2'\}, \{\}, \{\}\rangle\}\rangle.$$

Note that in merging the final case of (I_4) with the final case of (I_5), the two merged hypotheses are identical, and hence only one of them is retained. Since, for $\delta = \{x_1' \to x_2, y_1' \to y_2\}$, $\theta_1 = \{u \to y_1, v \to y_2\}$ and $\theta_2' = \{v \to y_1', w \to y_2'\}$, the substitutions $(\delta \circ \theta_1) \cup \gamma_2' = \{u \to y_1, v \to y_2, w \to y_2'\} = (\delta \circ \theta_2') \cup \gamma_1$, where $\gamma_1 = \{u \to y_1\}$ is the restriction of θ_1 on variables occurring only in (I_4) and $\gamma_2' = \{w \to y_2'\}$ is the restriction of θ_2' on variables occurring only in (I_5). Using the above merged cases, one can easily prove the conjecture (P_5). The above procedure for merging induction schemes preserves soundness and completeness as illustrated by the following theorem.

THEOREM 4. *An induction scheme ψ obtained by merging any two complete and sound basic induction schemes ϕ_f and ϕ_g that are mergeable is complete and sound.*

Proof. Completeness: Let ϕ_f and ϕ_g be respectively generated from the terms $t = f(x_1, \ldots, x_n)$ and $s = g(y_1, \ldots, y_k)$ of a conjecture $C(\ldots, t, \ldots, s, \ldots)$. Since the induction schemes ϕ_f and ϕ_g are complete, for any $n + k$-tuple $\langle e_1, \ldots, e_{n+k}\rangle$ of ground constructor terms, there exist induction cases: $\phi_f^i = \langle\langle\sigma_c, cond_c, repl_c\rangle, \{\ldots, \langle\theta_h, cond_h, repl_h\rangle, \ldots\}\rangle$ and $\phi_g^i = \langle\langle\sigma_c', cond_c', repl_c'\rangle, \{\ldots, \langle\theta_h', cond_h', repl_h'\rangle, \ldots\}\rangle$ such that for some ground substitutions γ and γ', $\gamma \circ \sigma_c(t) = e$, $\gamma(cond_c) = true$ and $\gamma' \circ \sigma_c'(s) = e'$, $\gamma(cond_c') = true$ where $e = f(e_1, \ldots, e_n)$ and $e' = g(e_{n+1}, \ldots, e_{n+k})$.

Without loss of generality, we assume ϕ_f^j and ϕ_g^j to be mergeable induction cases. Let $\beta = \gamma \cup \gamma'$; β agrees on its substitution on common variables in t and s. For a common variable $x_i = y_j$, $\sigma_c(x_i)$ and $\sigma_c'(y_j)$ are unifiable, $\beta(cond_c) = \beta(cond_c') = true$. The merged induction case $\psi_{(i,j)}$ generated from ϕ_f^i and ϕ_g^j, is $\langle\langle\sigma, mcond_c, mrepl_c\rangle, \{\ldots, \langle\beta_{(i,j)}, mcond_{(i,j)}, mrepl_{(i,j)}\rangle, \ldots\}\rangle$, where $\sigma = (\delta \circ \sigma_c) \cup (\delta \circ \sigma_c')$ with δ being the *mgu* of $\sigma_c(x)$ and $\sigma_c'(x)$, $x \in Vars(t) \cap Vars(s)$, and $mcond_c = \delta(cond_c) \cup \delta(cond_c')$. Thus $\beta = \theta \circ \delta$ for some substitution θ. Since $\theta \circ \sigma(t) = e$, and $\theta \circ \sigma(s) = e'$, and $\theta(mcond_c) = \theta \circ \delta(cond_c) \wedge \theta \circ \delta(cond_c') = true$, the $n + k$-tuple, $\langle e_1, \ldots, e_{n+k}\rangle$ is *covered* by the induction case $\psi_{(i,j)}$. Therefore, the induction scheme ψ is complete.

Soundness: By contradiction. Assuming a counterexample made of ground constructor terms serving as instances of t and s, a smaller counterexample is constructed in which either an instance of t or an instance of s is smaller. Assume that the conjecture $C(\ldots, t, \ldots, s, \ldots)$ is not true but the subgoal obtained from each induction case of ψ is true. Let $\langle e_1, \ldots, e_{n+k}\rangle$ be the smallest $n+k$-tuple of ground constructor terms, with respect to the ordering $\succ$ used to prove the termination of the definitions of g and f, that is a counterexample to C, i.e., $\theta \circ \sigma(C) = \mathit{false}$. The induction subgoal obtained from $\psi_{(i,j)}$, $(\sigma(C)\ \mathit{if}\ mcond_c)$ assuming $(\bigwedge_{(i,j)} \beta_{(i,j)}(C)\ \mathit{if}\ mcond_{(i,j)})$ is *true*. Since $\theta(\sigma(C)) = \mathit{false}$, and $\theta(mcond_c) = \mathit{true}$, one of $\theta(\beta_{(i,j)}(C))$ is *false*. But each $\beta_{(i,j)}$ (by definition mergeable induction cases) is either of the form $(\delta \circ \theta_h) \cup \theta_1$ or is of the form $(\delta \circ \theta'_h) \cup \theta_2$, both of which imply the existence of a counterexample $\langle e'_1, \ldots, e'_{n+k}\rangle$ to C which is lower in the ordering $\succ$ than $\langle e_1, \ldots, e_{n+k}\rangle$, where θ_1 and θ_2 are respectively the restrictions of θ'_h and θ_h on variables occurring only in them. By definition of cover set, $\sigma_c(t) \succ \theta_h(t)$ and $\sigma'_c(s) \succ \theta'_h(s)$ and for any hypothesis obtained from ϕ^i_f, $\theta(\sigma(t)) \succ \theta(\delta \circ \theta_h(t))$, i.e., $e \succ f(e'_1, \ldots, e'_n)$ and $\theta(\sigma(s)) \succ \theta(((\delta \circ \theta_h) \cup \theta_1)(s))$, since for any variable y of s that does not occur in t, $\theta(\sigma'_c(y)) \succ \theta(\theta'_h(y))$ and for a common variable y' of s and t, $\theta(\delta \circ \sigma'_c(y')) \succ (\delta \circ \theta_h(y'))$, since $\delta \circ \sigma'_c(y') = \delta \circ \sigma_c(y')$. Thus, $\langle e, e'\rangle \succ \langle f(e'_1, \ldots, e'_n), g(e'_{n+1}, \ldots, e'_{n+k})\rangle$ for the hypothesis obtained from ϕ^i_f. A similar result follows for the hypothesis obtains from ϕ^j_g by symmetry. A contradiction. □

3.1. Merging Using Linear Arithmetic

As discussed in the previous subsection, one of the crucial preconditions for merging two induction cases is to have syntactically unifiable substitutions for the common induction variables; the merged induction case is obtained from the *mgu* of these substitutions. If the definitions involved are not constructor based or if there are relations among the constructors (E is not empty), then the merged scheme obtained by the above described procedure using syntactic unification need not be complete. For such cases, the substitutions for the common induction variables in the induction cases have to be semantically unified (with respect to $=_E$).[12]

For example, in P_2: $divides(2, x) = not(divides(2, s(x)))$, the induction schemes suggested by $divides(2, x)$ and that suggested by $divides(2, s(x))$, given at the end of Section 2.4, cannot be merged by using the above procedure to obtain a complete and sound scheme because the substitution $\{x \to v+2\}$ in the scheme of $divides(2, x)$ is not unifiable with the substitution for x in any of cases of the scheme of $divides(2, s(x))$, and similarly the substitution $\{x \to v'+1\}$ in the latter scheme is irreconcilable with any of the cases of the former scheme.

However, these two induction schemes can be merged if we use linear arithmetic to reconcile these substitutions. The linear arithmetic procedure is again used to perform the semantic unification of the substitutions of the common induction variables among the schemes being merged. For soundness, the same semantic ordering $\succ_{LA}$ must be used to prove the termination of the definitions of the functions from which the schemes are derived.

We illustrate the merging algorithm by merging the scheme for $divides(2, x)$ with the scheme for $divides(2, s(x))$. The precise details are given in the algorithm in Subsection 3.1.1. The first element of the scheme for $divides(2, x)$ merges only with the first element of $divides(2, s(x))$ to give $\langle\langle\{x \to 0\}, \{\}, \{1 \leftarrow divides(2, 0), 2.1 \leftarrow divides(2, s(0))\}\rangle, \{\}\rangle$.

The second element of the scheme for $divides(2, x)$ merges only with the second element of $divides(2, s(x))$ to give

$$\langle\langle\{x \to 1\}, \{\}, \{1 \leftarrow divides(2, 1), 2.1 \leftarrow divides(2, 2 + 0)\}\rangle, \{\}\rangle.$$

Note that the hypothesis $\langle\{x \to v' - 1\}, \{v' \geqslant 1\}, \{2.1 \leftarrow divides(2, v')\}\rangle$ is discarded because the constraint equations of the induction conclusions of the two elements being merged are $\{x = 1, x = v' + 1\}$, which implies $v' = 0$, and the condition $v' \geqslant 1$ governing the hypothesis is unsatisfiable.

Finally, the third element of $divides(2, x)$ merges only with the second element of $divides(2, s(x))$ to give $\langle\langle\{x \to 2 + v\}, \{\}, \{1 \leftarrow divides(2, 2 + v), 2.1 \leftarrow divides(2, 2 + (v + 1))\}, \{\langle\{x \to v\}, \{\}, \{1 \leftarrow divides(2, v), 2.1 \leftarrow divides(2, v + 1)\}\rangle\}\rangle$. Note that to obtain the above merged induction scheme element, we reconcile the two substitutions of x: $x \to v + 2$ and $x \to v' + 1$, producing $v' = v + 1$.

As was shown in the Introduction, using the resulting induction scheme (I_3) one can easily prove the conjecture P_2.

3.1.1. *Linear Arithmetic-based Algorithm for Merging Induction Schemes*

Below we give an algorithm using LA for merging induction schemes that share common induction variables of number type. The algorithm can be generalized to consider any arbitrary E by using $=_E$ unification instead of semantic unification using LA. Of course, in order to do so, $=_E$ unification must be decidable and finitary, as mentioned before.

- **Input**: Induction schemes I_1 and I_2 that meet the following conditions.
  1. $indvars(I_1) \cap indvars(I_2) \neq \{\}$.
  2. The substitutions are over the theory of linear arithmetic for the common induction variables of I_1 and I_2.
- **Output**: An induction scheme I_{out}.
- **Method**:
  1. *Merge two distinct schemes I_1 and I_2*: For each induction case: $i_1 = \langle\langle\sigma_c, cond_c, repl_c\rangle, \{\ldots, \langle\theta_i, cond_i, repl_i\rangle, \ldots\}\rangle$ of I_1 do

a) For each induction case:
$i_2 = \langle\langle\sigma'_c, cond'_c, repl'_c\rangle, \{\ldots, \langle\theta'_j, cond'_j, repl'_j\rangle, \ldots\}\rangle$ of I_2 do

i) Let $CE = \sigma_c \cup \sigma'_c$ where σ_c and σ'_c are viewed as equations. Solve CE using LA with context $\{cond_c, cond'_c\}$. If solvable, let $\delta = \{x_i \to s_i \mid x_i \in Vars(\sigma_c) \cup Vars(\sigma'_c), x_i \notin s_i\}$ with feasibility constraints c_c. CE being unsolvable corresponds to no merged induction case being generated.

ii) Compute the merged induction scheme element $\langle\langle\sigma, mcond_c, mrepl_c\rangle, MH\rangle$, where

A) σ is the union of the substitutions $\delta \circ \sigma_c$ and $\delta \circ \sigma'_c$; $mcond_c$ is the union of $\delta(cond_c)$, $\delta(cond'_c)$ and $\delta(c_c)$; $mrepl_c$ is the union of the two replacements $\delta(repl_c)$ and $\delta(repl'_c)$.

B) For each triple $\langle\theta_i, cond_i, repl_i\rangle$ in the hypotheses of i_1, the triple $\langle(\delta\circ\theta_i)\cup\gamma'_j, \delta(cond_i)\cup\delta(cond'_j), \delta(repl_i)\rangle$ is included in MH if $(\delta \circ \theta_i) \cup \gamma'_j$ when viewed as equations is consistent with respect to LA with context $\{cond_i, cond'_j\}$. If the second component of i_2 is empty, then γ'_j is restriction of σ'_c on variables occurring only in i_2. Otherwise, γ'_j is the restriction of θ'_j on such variables.

Similarly, for each triple $\langle\theta'_j, cond'_j, repl'_j\rangle$ in the hypotheses of i_2, the triple $\langle(\delta \circ \theta'_j)\cup\gamma_i, \delta(cond'_j)\cup\delta(cond_i), \delta(repl'_j)\rangle$ is included in MH if $(\delta\circ\theta'_j)\cup\gamma_i$ when viewed as equations is consistent with respect to LA with context $\{cond'_i, cond'_j\}$. If the second component of i_1 is empty, then γ_i is the restriction of σ_c on variables occurring only in i_1. Otherwise, γ_i is the restriction of θ_i on such variables.

2. Return the set of merged cases to be the induction scheme I_{out}.

3.1.2. *Examples*

We illustrate the algorithm in an example based on the definition of gcd given earlier and the following definition of $even$ predicate on natural numbers.

1. $even(0) \to true$,
2. $even(1) \to false$,
3. $even(s(s(u))) \to even(u)$.

Consider proving the conjecture

$$(P_6)\colon\ gcd(m, 2) = 2 \text{ } if \text{ } even(m).$$

The scheme suggested by $even(m)$ at position 2.2 in (P_6) is

$$\{\langle\langle\{m \to 0\}\rangle, \{\}, \{2.2 \leftarrow even(0)\}\rangle, \{\}\rangle, \langle\langle\{m \to 1\}\rangle, \{\},$$
$$\{2.2 \leftarrow even(s(0))\}\rangle, \{\}\rangle,$$
$$\langle\langle\{m \to s(s(u))\}, \{\}, \{2.2 \leftarrow even(s(s(u)))\}\rangle,$$
$$\{\langle\{m \to u\rangle, \{\}, \{2.2 \leftarrow even(u)\}\rangle\}\rangle\},$$

and the scheme suggested by the $gcd(m, 2)$ at position 1 in (P_6) using LA is

$$\{\langle\langle\{m \to 0\}, \{\}, \{1 \leftarrow gcd(0, 2)\}\rangle, \{\}\rangle,$$
$$\langle\langle\{m \to 1\}, \{\}, \{1 \leftarrow gcd(1, 1 + 1)\}\rangle, \{\}\rangle,$$
$$\langle\langle\{m \to 2\}, \{\}, \{1 \leftarrow gcd(2, 2 + 0)\}\rangle, \{\}\rangle,$$
$$\langle\langle\{m \to x + 2\}, \{\}, \{1 \leftarrow gcd(x + 2, 2)\}\rangle,$$
$$\{\langle\{m \to x\}, \{\}, \{1 \leftarrow gcd(x, 2)\}\rangle\}\rangle\}.$$

Merging the scheme suggested by the subterm $even(m)$ into scheme suggested by the subterm $gcd(m, 2)$ is done as follows.

1. The first induction case of $even(m)$ merges, only with the first induction case of $gcd(m, 2)$. For the other cases, the constraint equations $\{0 = 2\}$, $\{0 = 1\}$ and $\{0 = x + 2\}$, respectively, are unsolvable. The merged case is $\langle\langle\{m \to 0\}, \{\}, \{1 \leftarrow gcd(0, 2), 2.2 \leftarrow even(0)\}\rangle, \{\}\rangle$.
2. The second induction case of $even(m)$ merges only with the second induction case of $gcd(m, 2)$, since the constraint equations are unsolvable for the rest of the cases. The merged case is $\langle\langle\{m \to 1\}, \{\}, \{1 \leftarrow gcd(1, 1 + 1), 2.2 \leftarrow even(1)\}\rangle, \{\}\rangle$.
3. For the third induction case of $even(m)$ with the third induction case of $gcd(m, 2)$ the constraint equations for the conclusion are $\{m = 2, m = s(s(u))\}$. The substitution obtained is $\{m \to 2, u \to 0\}$. The constraint equations for the hypothesis $\{m = u, u = 0\}$ are trivially solvable. The merged case obtained is $\langle\langle\{m \to 2\}, \{\}, \{1 \leftarrow gcd(2, 2 + 0), 2.2 \leftarrow even(2)\}\rangle, \{\langle\{m \to 0\}, \{\}, \{2.2 \leftarrow even(0)\}\rangle\}\rangle$.
4. For the third induction case of $even(m)$ with the final induction case of $gcd(m, 2)$, the constraint equations for the conclusion are $\{m = x + 2, m = s(s(u))\}$. The substitution obtained is $\{u \to x\}$. The constraint equations for the hypothesis are trivially solvable, and the merged induction case obtained is $\langle\langle\{m \to x + 2\}, \{\}, \{1 \leftarrow gcd(x + 2, 2), 2.2 \leftarrow even(s(s(x)))\}\rangle, \{\langle\{m \to x\}, \{\}, \{1 \leftarrow gcd(x, 2), 2.2 \leftarrow even(x)\}\rangle\}\rangle$.

Using the merged induction cases, one can easily prove the conjecture.

As the next example, consider the following conjecture from [11],

$$(P_7)\colon\ quot(m, 4) = hf(hf(m)),$$

based on the following two definitions: *quot*, which computes the quotient of dividing a natural number x by a natural number y, and hf, which halves a natural number,

$$1.\ quot(x,0) \to 0, \qquad 2.\ quot(x,y) \to 0\ if\,(x<y),$$
$$3.\ quot(x+y,y) \to s(quot(x,y))\ if\,(y \neq 0).$$
$$1.\ hf(0) \to 0, \qquad 2.\ hf(s(0)) \to 0, \qquad 3.\ hf(s(s(u))) \to s(hf(u)).$$

The induction scheme obtained from $hf(m)$ at position 2.1 in (P_7) is

$$\begin{array}{l}\{\langle\langle\{m \to 0\},\{\},\{2.1 \leftarrow hf(0)\}\rangle,\{\}\rangle,\\ \langle\langle\{m \to s(0)\},\{\},\{2.1 \leftarrow hf(s(0))\}\rangle,\{\}\rangle,\\ \langle\langle\{m \to s(s(u))\},\{\},\{2.1 \leftarrow hf(s(s(u)))\}\rangle,\\ \qquad \{\langle\{m \to u\},\{\},\{2.1 \leftarrow hf(u)\}\rangle\}\rangle\},\end{array}$$

and that obtained from $quot(m,4)$ at position 1 in (P_7) using LA is

$$\begin{array}{l}\{\langle\langle\{m \to 0\},\{\},\{1 \leftarrow quot(0,4)\}\rangle,\{\}\rangle,\\ \langle\langle\{m \to 1\},\{\},\{1 \leftarrow quot(1,4)\}\rangle,\{\}\rangle,\\ \langle\langle\{m \to 2\},\{\},\{1 \leftarrow quot(2,4)\}\rangle,\{\}\rangle,\\ \langle\langle\{m \to 3\},\{\},\{1 \leftarrow quot(3,4)\}\rangle,\{\}\rangle,\\ \langle\langle\{m \to x+4\},\{\},\{1 \leftarrow quot(x+4,4)\}\rangle,\\ \qquad \{\langle\{m \to x\},\{\},\{1 \leftarrow quot(x,4)\}\rangle\}\rangle\}.\end{array}$$

The scheme suggested by the subterm $hf(m)$ is merged into the scheme suggested by the subterm $quot(m,4)$ as follows.

1. The first two cases of $hf(m)$ merge only with the first two cases of $quot(m,4)$ resulting in $\langle\langle\{m \to 0\},\{\},\{\}\rangle,\{\}\rangle$ and $\langle\langle\{m \to s(0)\},\{\},\{\}\rangle,\{\}\rangle$, respectively.
2. The last case of $hf(m)$ merges with the third and the fourth cases of $quot(m,4)$, resulting in

   $$\langle\langle\{m \to 2\},\{\},\{1 \leftarrow quot(2,4), 2.1 \leftarrow hf(s(s(0)))\}\rangle,\langle\{m \to 0\},\{\},\ \{1 \leftarrow quot(0,4), 2.1 \leftarrow hf(0)\}\rangle\}\rangle,$$

   and

   $$\langle\langle\{m \to 3\},\{\},\{1 \leftarrow quot(3,4), 2.1 \leftarrow hf(s(s(s(0))))\}\rangle,\langle\{m \to 1\},\ \{\},\{1 \leftarrow quot(1,4), 2.1 \leftarrow hf(s(0))\}\rangle\}\rangle,$$

   respectively.
3. For the last case, of $hf(m)$ with the final case of $quot(m,4)$ the constraint equations for the conclusion are $\{m = s(s(u)), m = x+4\}$. The substitution obtained is $\{m \to x+4, u \to x+2\}$. The constraint equations set up for the two individual hypotheses are $\{m = x, u = x+2\}$, resulting in the substitution $\{m \to x\}$ and $\{m = u, u = x+2\}$, resulting in the substitution $\{m \to x+2\}$. Two hypotheses are generated corresponding to these two substitutions. The merged induction case generated is

$$\{\langle\langle\{m \to x+4\}, \{\}, \{1 \leftarrow quot(x+4,4), 2.1 \leftarrow hf(s(s(x+2)))\}\rangle,$$
$$\{\langle\{m \to x\}, \{\}, \{1 \leftarrow quot(x,4), 2.1 \leftarrow hf(x)\}\rangle,$$
$$\langle\{m \to x+2\}, \{\}, \{1 \leftarrow quot(x+2,4), 2.1 \leftarrow hf(x+2)\}\rangle\}.$$

As should be evident from the induction step, there is an induction hypothesis being contributed by each of the induction schemes corresponding to $quot(m, 4)$ and $hf(m)$. Using the above merged induction cases, one can easily prove the conjecture.

3.1.3. *Completeness and Soundness of a Merged Induction Scheme*

The completeness and soundness of the induction scheme generated by the above algorithm for merging sound and complete induction schemes by using LA for semantic unification are established in the same way as was done earlier for completeness and soundness of merging of basic induction schemes. Instead of unification over $=_E$, unification modulo LA is done. For each individual induction scheme, the proof is patterned after that of Theorem 3, establishing completeness and soundness of an induction scheme generated by using LA for semantic unification. So the following proof combines proofs of Theorems 3 and 4.

THEOREM 5. *The induction scheme ψ, obtained by merging any two sound and complete induction schemes ϕ_f and ϕ_g corresponding to complete LA-based cover sets C_f and C_g, respectively, is complete and sound.*

Proof. Completeness: Let ϕ_f and ϕ_g be respectively generated from $t = f(t_1, \ldots, t_n)$ and $s = g(s_1, \ldots, s_k)$ of a conjecture $C(\ldots, t, \ldots, s, \ldots)$. For any $n + k$-tuple $\langle e_1, \ldots, e_{n+k}\rangle$ of numbers such that for some ground substitutions, β and β', $\beta(t) =_{LA} e$ and $\beta'(s) = e'$, where $e = f(e_1, \ldots, e_n)$ and $e' = g(e_{n+1}, \ldots, e_{n+k})$, there exist induction cases $\phi_f^i = \langle\langle \sigma_c, cond_c, repl_c\rangle, \{\ldots, \langle\theta_h, cond_h, repl_h\rangle, \ldots\}\rangle$ and

$$\phi_g^j = \langle\langle \sigma'_c, cond'_c, repl'_c\rangle, \{\ldots, \langle\theta'_h, cond'_h, repl'_h\rangle, \ldots\}\rangle$$

which *cover* $\langle e_1, \ldots, e_{n+k}\rangle$. Thus, for some ground substitutions γ and γ', $\gamma \circ \sigma_c(t) =_{LA} e$, $\gamma(cond_c) = true$ and $\gamma' \circ \sigma'_c(s) =_{LA} e'$, $\gamma'(cond'_c) = true$.

Let $\beta'' = \gamma \cup \gamma'$; β'' agrees on its substitution on common variables in t and s. For a common variable x, $\sigma_c(x)$ and $\sigma'_c(x)$ are unifiable modulo LA. $\beta''(cond_c) = \beta''(cond'_c) = true$. It follows along the same lines as Theorem 4 that the $n + k$-tuple, $\langle e_1, \ldots, e_{n+k}\rangle$ is *covered* by the induction case, $\psi_{(i,j)}$, generated from ϕ_f^i and ϕ_g^j by Algorithm 3.1.1, and therefore the induction scheme ψ is complete.

Soundness: By contradiction. Assuming a counterexample of numbers serving as instances of t and s, a smaller counterexample is constructed in the same manner as in Theorem 4, in which either an instance of t or an instance of s is smaller. Let $\langle e_1, \ldots, e_{n+k}\rangle$ be the smallest $n+k$-tuple of numbers, with respect to the ordering $\succ_{LA}$ used to prove the termination of the definitions of g and f, that is a counterexample to C. Since $\theta(\sigma(C)) = \mathit{false}$, and $\theta(mcond_c) = \mathit{true}$, one of $\theta(\beta_{(i,j)}(C))$ is *false*. But each $\beta_{(i,j)}$ computed by the algorithm in Section 3.1.1 is either of the form $(\delta \circ \theta_h) \cup \theta_1$ or of the form $(\delta \circ \theta'_h) \cup \theta_2$, both of which imply the existence of a counterexample $\langle e'_1, \ldots, e'_{n+k} \rangle$ of numbers to C, which is lower in the ordering $\succ_{LA}$ than $\langle e_1, \ldots, e_{n+k} \rangle$, where θ_1 and θ_2 are respectively the restrictions of θ'_h and θ_h on variables occurring only in them. A contradiction. □

4. Checking Completeness of Cover Sets

An important property of a cover set is that it be a complete cover for the data structure under consideration. As proved earlier, an induction scheme generated by using a complete cover set is sound. This is ensured if the terminating rewrite rules defining a function whose left sides are used as the basis for constructing the cover set completely define the function. In [5, 8], algorithms for checking completeness of constructor-based definitions expressed as terminating rewrite rules are discussed; for background information on this topic, an interested reader may also consult [6]. For LA-based definitions the linear arithmetic procedure can be used for checking the completeness. This fills a major gap in the use of the cover set method for mechanizing induction in *RRL*. For example, in the proof of unique prime factorization theorem reported in [13], the completeness of the definitions of some of the functions (such as div, rem, gcd, $primefac$, etc.) and the associated cover sets was established manually. With the linear arithmetic procedure, all these proofs can be carried out automatically in *RRL*. We develop an algorithm for checking completeness of function definitions and the associated cover sets using the linear arithmetic procedure. We focus on function definitions in which all arguments are of number type; the method can be easily extended to consider definitions taking arguments from numbers as well as other data structures such as lists and sequences. This algorithm serves as a decision procedure for checking the completeness of LA-based cover sets and thus could be used as a heuristic for checking the completeness of LA-based function definitions.

4.1. Algorithm to Check Completeness of a Function Definition

- **Input**: An LA-based cover set C_f of an n-ary function $f(x_1, \ldots, x_n)$.
- **Output**: A quantifier-free linear arithmetic formula ξ with free variables $x_1, \ldots, x_n$, specifying the values over which f is defined; true if C_f is complete.
- **Method**:
  1. *Initialize*: Initialize ξ to be *false*.
  2. *Linear Solve*: For each cover set triple
     $c_i = \langle \langle s_1, \ldots, s_n \rangle, \{\ldots, \langle s'_1, \ldots, s'_n \rangle, \ldots\}, cond \rangle$ do:

a) Set up constraint equations $CE = \{x_i = s_i \mid 1 \leqslant i \leqslant n\}$.
b) Solve CE by LA with context *cond*. If solvable, let c_c be a quantifier-free formula over x_i's under which solutions exist. If no solution exists, then c_c is assumed to be false.
c) $\xi := \xi \vee c_c$.

3. *Linear Simplify*: Using LA, check whether ξ is valid. If so, then return *true*. Otherwise, return the simplified form of ξ.

As the following theorem shows, if the above algorithm returns *true*, then the definition of f is complete, and hence the associated cover set is also complete. If the algorithm returns a formula different from *true*, then the definition of f may or may not be complete depending upon whether the rewrite rules defining f are confluent or not. However, the associated cover set is incomplete.

THEOREM 6. *An LA-based cover set C_f of a function $f(x_1, \ldots, x_n)$ is complete iff the above algorithm returns true.*

Proof. If an LA-based cover set C_f of an n-ary function f is complete, for any n-tuple $\langle e_1, \ldots, e_n \rangle$ of numbers, there exist a substitution σ and a cover set triple $c_i = \langle \langle s_1, \ldots, s_n \rangle, \{\ldots, \langle s'_1, \ldots, s'_n \rangle, \ldots\}, cond \rangle$ such that $\sigma(s_i) =_{LA} e_i$, $1 \leqslant i \leqslant n$, and $\sigma(cond) = true$. So the formula c_c generated by the algorithm for that cover set tuple at step 2(b) when instantiated with $\{x_1 \to e_1, \ldots, x_n \to e_n\}$ is *true*, making $\xi[x_1 \to e_1, \ldots, x_n \to e_n]$ to be true. This implies that the quantifier-free formula ξ with free variables $x_1, \ldots, x_n$ generated by the algorithm is equivalent to *true*.

If C_f is not complete, there exists an n-tuple $\langle e_1, \ldots, e_n \rangle$ that is not covered by any cover set triple; i.e., for each cover set triple c_i, the set of equations $CE = \{e_i = s_i\}$ in step 2(a) of the above algorithm does not have a solution under *cond*; i.e., the corresponding $c_c[x_1 \to e_1, \ldots, x_n \to e_n]$ is false. Therefore, $\xi[x_1 \to e_1, \ldots, x_n \to e_n]$ is false, implying that ξ cannot be equivalent to *true*. □

4.2. EXAMPLES

- As the first example, consider the definition of gcd given earlier.
  1. *Initialize*: Let $\xi = false$ and C_{gcd} be the cover set of $gcd(x_1, x_2)$.
  2. *Linear Solve*: For the first two cover set triples of C_{gcd} the constraint equations are $\{x_1 = x, 0 = x_2\}$ and $\{x_1 = 0, x_2 = x\}$, respectively. Hence ξ is updated to be $\xi = (x_1 = 0) \vee (x_2 = 0)$.
     For the third cover set triple, the constraint equations are $\{x + y = x_1, y = x_2\}$ with context $x + y > x$. c_c is $x_1 \geqslant x_2 \wedge x_2 > 0$. Adding these conditions, we obtain $\xi = (x_1 = 0) \vee (x_2 = 0) \vee (x_1 \geqslant x_2 \wedge x_2 > 0)$.
     For the last cover set triple by symmetry, c_c is $x_2 \geqslant x_1 \wedge x_1 > 0$. Adding these two conditions to ξ, we obtain

$$\xi = (x_1 = 0) \vee (x_2 = 0) \vee (x_1 \geqslant x_2 \wedge x_2 > 0) \\ \vee (x_2 \geqslant x_1 \wedge x_1 > 0).$$

3. *Linear Simplify*: Since LA reduces ξ to *true*, the definition of *gcd* is complete

– *Integer Divisibility*: Consider the predicate *divides* defined in the example discussed in the Introduction. We observed that $divides(x_1, x_2)$ is defined only if its first argument is non-zero or both its arguments are 0. This can be established by using the above algorithm as follows.
  1. *Initialize*: Let $\xi = false$, and $C_{divides}$ be the cover set for *divides*.
  2. *Linear Solve*: For the first cover set triple, the constraint equations are $\{x_1 = u, x_2 = 0\}$. c_c is $\{x_2 = 0\}$. So, $\xi = (x_2 = 0)$.
  With the second cover set triple $\langle\langle u, v\rangle, \{\}, \{v < u, v \neq 0\}\rangle$, the constraint equations are $\{x_1 = u, x_2 = v\}$, with context $\{v < u, v \neq 0\}$. c_c is $x_2 < x_1 \wedge x_2 > 0$, so $\xi = (x_2 = 0) \vee (x_2 < x_1 \wedge x_2 > 0)$.
  For the third cover set triple $\langle\langle u, u + v\rangle, \{\langle u, v\rangle\}, \{u + v > v\}\rangle$, the constraint equations are $\{x_1 = u, x_2 = u + v\}$ with context $u + v > v$. c_c is $x_2 \geqslant x_1 \wedge x_1 > 0$. The updated ξ is

$$\xi = (x_2 = 0) \vee (x_2 < x_1 \wedge x_2 > 0) \vee (x_2 \geqslant x_1 \wedge x_1 > 0).$$

  3. *Linear Simplify*: Since ξ is not a valid formula, LA does not reduce it to *true*. The simplified form of ξ, $x_2 = 0 \vee x_1 > 0$, is output by the algorithm, which corresponds to our initial observation that *divides* is defined only if when its first argument is non-zero or the second argument is zero.

4.2.1. *Incomplete Cover Sets*

In order to draw sound conclusions, it is important to use a complete cover set. Otherwise, unsound conclusions can be made. For example, if the cover set based on the definition of *divides* given in the Introduction is used without caring about its completeness, it is possible to conclude wrongly that

$$divides(x, y + y) \textit{ if } divides(x, y),$$

even though the above formula does not hold for the case when $x = 0$. Since *divides* is defined only if $x > 0 \vee y = 0$, the above formula holds only under this condition. Thus,

$$divides(x, y + y) \textit{ if } divides(x, y) \wedge (x > 0 \vee y = 0)$$

is a theorem. Similarly, in the case of the second conjecture,

$$divides(2, x) = not(divides(2, s(x))) \textit{ if } (2 > 0 \vee x = 0)$$

is a theorem, which is the same as

$$divides(2, x) = not(divides(2, s(x))),$$

since the condition in the conditional formula is true.

Incomplete cover sets, however, can be useful in certain cases, such as proving $divides(2, x) = not(divides(2, s(x)))$. An incomplete cover set of *divides* can be used since the cover set is incomplete only if the first argument of *divides* is 0. We can relativize the completeness of a cover set with respect to a formula.

DEFINITION. *Relatively complete LA-based cover sets*: An LA-based cover set C_f of $f(x_1, \ldots, x_n)$ is *complete with respect to* a linear arithmetic quantifier-free formula ξ with free variables $x_1, \ldots, x_n$ iff for any n-tuple $\langle e_1, \ldots, e_n\rangle$ of numbers either one of the following conditions hold.

- $\xi[x_1 \to e_1, \ldots, x_n \to e_n] = false$,
- there exists a cover set triple $c_i = \langle\langle s_1, \ldots, s_n\rangle, \{\ldots, \langle s'_1, \ldots, s'_n\rangle, \ldots\}, cond\rangle$ and a substitution σ such that $\sigma(s_i) =_{LA} e_i$ and $\sigma(cond) = true$.

A complete LA-based cover set C_f is complete with respect to the formula *true*. A cover set C_f associated with $f(x_1, \ldots, x_n)$ can be used to establish an inductive property C *if* $\xi(x_1, \ldots, x_n)$ if C_f is complete with respect to $\xi(x_1, \ldots, x_n)$.

THEOREM 7. *Given an LA-based cover set C_f of $t = f(x_1, \ldots, x_n)$ that is complete with respect to $\xi(x_1, \ldots, x_n)$, an induction scheme generated from C_f is sound and complete for a conjecture $C(\ldots, t, \ldots)$ if $\xi(x_1, \ldots, x_n)$.*

Proof. By contradiction. Assume that $C(\ldots, t, \ldots)$ *if* $\xi(x_1, \ldots, x_n)$ is not true, but for each induction case of a scheme ψ obtained from C_f, $C(\ldots, t, \ldots)$ *if* $\xi(x_1, \ldots, x_n)$ is true. Let $\langle e_1, \ldots, e_n\rangle$ be the smallest n-tuple of numbers, with respect to the termination ordering $\succ_{LA}$ used to prove termination of the definition of f, that is a counterexample to $C(\ldots, t \ldots)$ *if* $\xi(x_1, \ldots, x_n)$. That is, $C(\ldots, f(x_1, \ldots, x_n), \ldots)[x_1 \to e_1, \ldots, x_n \to e_n]$ *if* $\xi[x_1 \to e_1, \ldots, x_n \to e_n] = false$. Equivalently,

$$\xi[x_1 \to e_1, \ldots, x_n \to e_n] = true$$

and

$$C(\ldots, f(x_1, \ldots, x_n), \ldots)[x_1 \to e_1, \ldots, x_n \to e_n] = false\,.$$

Since C_f is complete with respect to $\xi(x_1, \ldots, x_n)$ and $\xi[x_1 \to e_1, \ldots, x_n \to e_n] = true$, there exist a cover set triple,

$$c_i = \langle\langle s_1, \ldots, s_n\rangle, \{\ldots, \langle s_1^k, \ldots, s_m^k\rangle, \ldots\}, cond\rangle,$$

and a substitution σ such that $\sigma(s_i) =_{LA} e_i$, $1 \leqslant i \leqslant n$, and $\sigma(cond) = true$. Corresponding to c_i, there exists an induction case ψ_j of the form, $\langle\langle\{x_1 \to$

$s_1, \ldots, x_n \to s_n\}, cond, \{\}\rangle, \{\ldots, \langle\{x_1 \to s_1^k, \ldots, x_n \to s_n^k\}, cond, \{\}\rangle, \ldots\}\rangle$, by Theorem 2.6. The induction subgoal corresponding to ψ_j, $(\theta(C)\ if\ \theta(\xi))\ if\ cond$ assuming $\{\bigwedge_k(\theta_k(C)\ if\ \theta_k(\xi))\ if\ cond\}$ is *true*, where $\theta = \{x_1 \to s_1, \ldots, x_n \to s_n\}$ and $\theta_k = \{x_1 \to s_1^k, \ldots, x_n \to s_n^k\}$. Since $\sigma(\theta(C)\ if\ \theta(\xi))$ is *false* and $\sigma(cond) = true$, one of $\sigma((\theta_k(C)\ if\ \theta_k(\xi))\ if\ cond)\}$ is false. $\sigma(cond)$ being *true* implies that there is a smaller counterexample to $C\ if\ \xi$ since $\sigma(\langle s_1^k, \ldots, s_n^k\rangle)$ is lower in the termination ordering $\succ_{LA}$ than $\langle e_1, \ldots, e_n\rangle$ (by the definition of an LA-based cover set, $\langle s_1, \ldots, s_n\rangle \succ_{LA} \langle s_1^k, \ldots, s_n^k\rangle$ for any k, and since $\succ_{LA}$ is preserved under $=_{LA}$, $\langle\sigma(s_1), \ldots, \sigma(s_n)\rangle \succ_{LA} \langle\sigma(s_1^k), \ldots, \sigma(s_n^k)\rangle$). A contradiction. □

5. Generalization

While attempting proofs by induction, intermediate conjectures are generated that may be difficult to prove automatically. In some situations, however, it is possible to prove a generalization of a conjecture from which the conjecture follows. Most induction theorem provers support heuristics for generalizing conjectures. In *RRL*, at least two kinds of generalizations are performed: (i) abstracting a nonvariable subterm appearing in a conjecture to be a variable, if the nonvariable subterm appears more than once in the conjecture (in the left side as well as the right side and/or the condition), and (ii) dropping an assumption from a conditional conjecture. Semantic analysis can be useful in the implementation of the generalization heuristic, since a subterm may have multiple occurrences semantically but it may not appear to have multiple occurrences syntactically. In this section, we illustrate the use of the linear arithmetic procedure for improving the generalization heuristic focusing on the first kind of generalization of abstracting a nonvariable subterm with multiple occurrences by a variable. We first briefly review how this generalization heuristic is performed in *RRL* based on syntactic properties without using LA, and then we discuss how this heuristic can be improved using LA.

Given a conjecture C of the form $l = r\ if\ cond$, we look for a maximal nonvariable subterm s occurring in at least two of l, r and $cond$. Since there may be many such maximal subterms, they are collected in a list as possible candidates for generalization. Abstracting any nonempty subset of a set of candidates would lead to a generalized version of the conjecture. For each such subset of candidates, a *generalization template* is generated in which the first component is a list of *abstraction pairs* of the form $\langle s, u\rangle$, where s is a candidate subterm and u is a new variable used to replace s, and the second component is a list of triples of the form $\langle p, t_1, t_2\rangle$ indicating that subterm t_1 at position p in C is replaced by t_2 using the abstraction pairs (obviously, t_2 is generated from t_1 by abstracting using the first component). A generalization template is used to generate a generalized version C_g of C by simultaneously replacing the subterm t_1 at position p in C

by t_2 for every triple $\langle p, t_1, t_2 \rangle$ in its second component. It is obvious that if C_g can be proved, then C can be proved.

The main steps of the generalization procedure in *RRL* are to (i) identify maximal nonvariable subterms that occur in a conjecture, (ii) generate different possible generalization templates by considering all possible nonempty lists of abstraction pairs, and finally (iii) use each generalization template, one by one, to generate a generalized conjecture, until a proof is obtained. Generation of candidate subterms for generalization and the order of replacement of candidate subterms by new variables are arbitrary and have been fine-tuned in *RRL* based on experimentation.

5.1. Using Linear Arithmetic for Determining Subterms

A drawback of the above heuristic, as was the case with other heuristic discussed earlier, is that it is based on syntactic considerations only. A subterm may not have multiple occurrences syntactically in a conjecture, but semantically equivalent subterms may occur in the conjecture. For instance, consider the following conjecture about the *gcd* function:

$$gcd(x + y + 1, 2 * x + 2 * y + 2) = x + y + 1.$$

The first argument of *gcd* appears to be quite different from the second argument. If no semantic analysis is performed, *RRL* we would generate

$$gcd(u, 2 * x + 2 * y + 2) = u,$$

as a generalized version of the conjecture by abstracting occurrences of $x + y + 1$ in the left and right side by a variable u, since the second argument of gcd syntactically appears not to include $x + y + 1$ as a subterm. This generalization is not a valid formula because there are counterexamples. Even if the above subterms are expressed in some normal form, say $s(x + y)$ and $s(s(x + x + y + y))$, respectively, the first does not occur in the second. However, from a semantic standpoint, the second argument of gcd is closely related to its first argument. If LA is used to look for equivalent subterms, then this relationship can be identified, and it can be found that the first argument of gcd appears twice in its second argument.

Given a conjecture of the form $l = r\ if\ cond$, for some maximal nonvariable subterm s of l, we check whether it appears in r and/or $cond$ also. If s is not a linear term, syntactic subterm check is performed as before. Otherwise, we check whether other linear subterms include occurrences of s. This check can be performed using LA by querying whether a linear subterm $t \geqslant s$. If the answer is no, then s does not occur in t; otherwise, we find the number of times s appears in t (this can be done by repeated query and subtraction from t until the result becomes smaller than s). Let $t = k * s + tr$, where k is a positive integer and tr is a linear term smaller than s. In this case, s can be abstracted to be a new variable u giving the abstraction pair $\langle s, u\rangle$, and the subterm t in the conjecture at position p can be replaced by $k * u + tr$.

Once candidate subterms in a conjecture C for generalization have been identified, generalization templates can be generated by considering all possible subsets of candidates for generalization as before.

In the above conjecture about gcd, we look for occurrences of the linear term $x + y + 1$ in other linear terms in the conjecture. LA is queried to compute the number of occurrences of $x + y + 1$ in the other linear term $2 * x + 2 * y + 2$ in the conjecture. By repeated subtraction the number of occurrences is computed to be 2 with a remainder 0. Similarly, the right side of the conjecture also has an occurrence of $x + y + 1$. And, that is the only maximal subterm in the conjecture that can be generalized, since subterm $2*x+2*y+2$ does not occur in the other linear subterm $x + y + 1$. By abstracting $x + y + 1 = u$, the template formed is

$$\langle\{(x + y + 1, u)\}, \{\langle 1.1, x + y + 1, u\rangle, \langle 1.2, 2 * x + 2 * y + 2, 2 * u\rangle, \\ \langle 2, x + y + 1, u\rangle\}\rangle.$$

From this, the conjecture is generalized to $gcd(u, 2 * u) = u$. This conjecture can be rewritten by using the fourth rule of gcd to give $gcd(u, u) = u$, which can be proved using the LA-based induction scheme generation procedure, as shown earlier. A complete description of the extended generalization procedure using LA follows.

- **Input**: A conjecture C of the form $l = r$ *if cond*.
- **Output**: A set of generalized versions of conjecture C.
- **Method**: For each maximal nonvariable subterm s of C at position p do:
  1. *Generate Templates*:
     a) Let (s, x), $x \notin Vars(C)$ be the abstraction pair associated with s.
     b) *Compute the replacement triples*: Let the set of triples $S = \{\}$. For each nonvariable subterm t of C at a position q other than p do:
        i) If s is not a linear term and $t = s$, then $S := S \cup \{\langle q, s, x\rangle\}$.
        ii) If s is a linear term, then compute the number of occurrences, n of s in t, by using LA by repeated subtraction of s from t until $t \geqslant s$ with rt being the remainder, if any.
        iii) If $n = 0$, then no replacement triple is generated. Otherwise, $S := S \cup \{\langle q, t, nx + rt\rangle\}$.
  2. Templates whose second components do not replace terms occurring in at least two of l, r or *cond* are discarded.
  3. For each nonempty subset of templates, output the generalized version of C obtained by simultaneously replacing for each of the triples $\langle p, t_1, t_2\rangle$, the subterm t_1 at position p by t_2 in C.

6. Related Work

The seminal work on mechanizing induction is that of Boyer and Moore [1, 2]. In [1], Boyer and Moore discussed recursion analysis of definitions and the formulation of induction schemes for a given conjecture. Many other heuristics that manipulate induction schemes towards the choice of the most appropriate induction scheme for a given conjecture are also developed there. These methods have been implemented in their theorem prover *Nqthm*. All of these methods are purely syntactic and do not exploit the semantic information associated with the functions in any manner. For instance, in formulating an induction scheme, for a given conjecture based on a term $f(t_1, \ldots, t_n)$ they write [1, pp. 185–186].

> "If the t_i occupying the argument positions measured by the induction template we have in mind are all distinct variables, then we can use the template to construct a sound induction scheme. And in general, if any of the measured t_i is not a variable or is the same variable as another measured t_i we cannot. ... Let *changeables* be those t_i that occurred in measured positions that are also sometimes changed in the recursion. Let the *unchangeables* be the variables occurring in those t_i occupying measured positions that never change in recursion. A template *applies* to a term if the changeables are all distinct variables and none of them is among the unchangeables."

In almost all the examples discussed in this paper, the methods given in [1] would be inapplicable. For instance, in the definition of gcd both the arguments of gcd are required to justify the termination of the definition and hence are in measured positions. Further, both of them are changed in the recursive calls to gcd in the definition and are therefore, *changeables* in the sense of [1]. Thus, given a conjecture of the form $gcd(m+m, 2) = 2$, an induction scheme cannot be formulated corresponding to the term $gcd(m+m, 2)$, since the arguments to gcd are nonvariables. As yet another example, in the conjecture such as $gcd(m, m) = m$, an induction scheme cannot be formulated corresponding to the term $gcd(m, m)$ also since the arguments in changeable positions of gcd are not distinct variables. Similarly, in our motivating example, $divides(2, x) = not(divides(2, s(x)))$, $s(x)$ is a nonvariable in a changeable position of $divides$ and is not syntactically compatible with $+$ used in defining $divides$.

Nqthm provides a mechanism for performing user-directed induction by using the so-called *hint* directive [2]. A user can specify a function definition to be used for generating an induction scheme. The variables in the conjecture and additional dummy variables, if required, are provided as arguments to the function being specified. For the conjecture (P_2),

$$divides(2, x) = not(divides(2, s(x))),$$

the scheme generated by using LA can be obtained in *Nqthm* by using the hint directive

$$divides(2, x) = not(divides(2, s(x))) \text{ } if \text{ } induct(even(m)).$$[13]

This shows the power of the LA-based induction scheme generation procedure. Some induction schemes that must otherwise be specified by using *hint* can be automatically generated by using the linear arithmetic procedure.

The description of merging schemes in this paper follows along the lines of [1]. In [1] whenever two schemes are merged, the basis case of the merged scheme is obtained by negating the conditions of the merged induction cases. This is not the case in *RRL*, where the basis cases are generated from the rules that do not have recursive calls on the right-hand side. Consequently, for soundness, all induction cases have to be considered while merging.

7. Concluding Remarks

We have shown how a decision procedure such as Presburger arithmetic, which embodies knowledge about numbers and their two related representations using 0, s, $+$, can be effectively used to further automate and make more effective certain heuristics in mechanizing induction. The focus of this paper has been on generating induction schemes by using the decision procedure. We have briefly reviewed how the decision procedure can be used for checking completeness of definitions and hence the associated cover sets and induction schemes. This extension of the cover set method, which does semantic analysis with the help of the linear arithmetic decision procedure, enables the automation of proofs of many theorems that otherwise require human guidance and hints. We have manually exercised this extension quite a bit on a number of examples, and the results are extremely encouraging. Since *RRL* already supports a decision procedure for linear arithmetic [9], we plan to extend the procedure to handle the various applications discussed in this paper. We believe that the use of the linear arithmetic procedure for performing semantic analysis will also be useful in the proof by consistency (also popularly known as the inductionless induction) approach.

The proposed approach suggests a new direction for enhancing heuristics for mechanizing induction based on semantic information about data structures. We believe that in data structures such as finite sets and multisets, constructors such as the null element, inserting a single element, and the union operation play a role similar to the constructors 0, s, and $+$ for numbers. So it should be possible to design decision procedures and heuristics for converting between different representations of values of the data structure. This might be the case for other data structures also, such as lists and sequences. There is a need to study and investigate developing decision procedures for these data structures, especially focusing on aspects related to convertibility among representations.

Notes

[1] See, however, a remark later. If semantic information such as linear arithmetic is used for rewriting as well, then rule 3 is not terminating, for example, when u is 0. The termination condition required is $u + v > v$.

[2] Since *divides* is not defined when its first argument is 0 and its second argument is non-zero, the above conjecture is not true if we drop the condition $x > 0$ from it. The completeness of the definition of *divides* is discussed in detail in Section 4.

[3] A position is a sequence of non-negative integers used to refer to a subterm in a term. An equation will be considered as a term with $=$ as the binary predicate; a conditional equation will be considered as a term with $=$ as the binary predicate whose second argument is an if term, where if is considered as a binary function. In the above example, the position of $divides(x, y)$ is 2.2.1 as the conjecture is viewed as an abbreviation for

$$divides(x, y + y) = true \text{ if } divides(x, y) \wedge x > 0.$$

[4] This example is taken from the *Nqthm* corpus. This conjecture is proved there with the help of an explicit induction hint.

[5] Otherwise, it may be necessary to rewrite backwards, which would have to be done with a user providing guidance.

[6] $C[\{p_1 \leftarrow t_1, \ldots, p_n \leftarrow t_n\}]$ denotes the conjecture C with its subterm at position p_i replaced by the term t_i, $1 \leqslant i \leqslant n$.

[7] See, however, Subsection 4.1.1 in which it is shown how incomplete cover sets can be used to generate induction schemes that can also be useful in proofs by induction.

[8] Note that $v \geqslant 0$ is an implicit feasibility constraint for any natural number v, which is left unspecified.

[9] $\sigma(\{s_1, \ldots, s_k\}) = \{\sigma(s_1), \ldots, \sigma(s_k)\}$.

[10] Since the definitions are constructor based, a complete induction scheme can be generated from the non-basic term $same(cons(x_1, y_1), w)$, and a further induction based on this scheme could alternatively be used to establish the conjecture (P_5).

[11] It is assumed that the same reduction ordering is used for proving the termination of the two definitions, the induction schemes corresponding to which are being merged. Otherwise, the merged induction scheme need not be sound, since the hypotheses of both the schemes being merged are included in the hypotheses of the merged schemes.

[12] This can be done provided unification modulo $=_E$ is decidable and finitary. Further, merged induction cases corresponding to each *mgu* of the substitutions of the common induction variables need to be generated for the merged scheme to be complete.

[13] In the *Nqthm* corpus, an induction scheme based on the definition of the function *odd* is used instead.

References

1. Boyer, R. S. and Moore, J S.: *A Computational Logic*, ACM Monographs in Computer Science, 1979.
2. Boyer, R. S. and Moore, J S.: *A Computational Logic Handbook*, Academic Press, New York, 1988.
3. Boyer, R. S. and Moore, J S.: Integrating decision procedures into heuristic theorem provers: A case study of linear arithmetic, *Machine Intelligence* **11** (1988), 83–157.
4. Dershowitz, N.: Termination of rewriting, *J. Symbolic Computation* **3** (1987), 69–116.
5. Jouannaud, J.-P. and Kounalis, E.: Automatic proofs by induction in theories without constructors, *Information and Computation* **82** (1989), 1–33.
6. Kapur, D.: An automated tool for analyzing completeness of equational specifications, in *Proc. Int. Symp. Software Testing and Analysis (ISSTA)*, Seattle, August 1994, pp. 28–43.
7. Kapur, D., Musser, D. R., and Nie, X.: An overview of the Tecton proof system, in *Theoret. Computer Science Journal*, special issue on *Formal Methods in Databases and Software Engineering* (ed. V. Alagar), Vol. 133, October 1994, pp. 307–339.

8. Kapur, D., Narendran, P., Rosenkrantz, D., and Zhang, H.: Sufficient-completeness, quasi-reducibility and their complexity, *Acta Informatica* **28** (1991), 311–350.
9. Kapur, D. and Nie, X.: Reasoning about numbers in Tecton, in *Proc. 8th Int. Symp. Methodologies for Intelligent Systems (ISMIS'94)*, Charlotte, NC, October 1994, pp. 57–70.
10. Kapur, D. and Zhang, H.: An overview of Rewrite Rule Laboratory (RRL), *J. Computer Math. Appl.* **29**(2) (1995), 91–114.
11. Walther, C.: Combining induction axioms by machine, *Proc. 12th Int. Joint Conf. Artificial Intelligence*, Chambery, France, 1993.
12. Zhang, H.: Reduction, superposition and induction: Automated reasoning in an equational logic, Ph.D. Thesis, Department of Computer Science, Rensselaer Polytechnic Institute, Troy, NY, 1988.
13. Zhang, H., Kapur, D., and Krishnamoorthy, M. S.: A mechanizable induction principle for equational specifications, in *Proc. 9th Int. Conf. Automated Deduction (CADE-9)*, Argonne, IL, LNCS 310, Springer-Verlag, 1988, pp. 250–265.

Journal of Automated Reasoning **16:** 79–111, 1996.

Productive Use of Failure in Inductive Proof*

ANDREW IRELAND** and ALAN BUNDY
Department of Artificial Intelligence, University of Edinburgh, 80 South Bridge, EH1 1HN Edinburgh, Scotland, UK
e-mail: air@cee.hw.ac.uk, a.bundy@ed.ac.uk

Abstract. Proof by mathematical induction gives rise to various kinds of eureka steps, e.g., missing lemmata and generalization. Most inductive theorem provers rely upon user intervention in supplying the required eureka steps. In contrast, we present a novel theorem-proving architecture for supporting the automatic discovery of eureka steps. We build upon *rippling*, a search control heuristic designed for inductive reasoning. We show how the failure if rippling can be used in bridging gaps in the search for inductive proofs.

Key words: Automated theorem proving, mathematical induction, proof patching.

1. Introduction

1.1. Motivation

G. H. Hardy [12] draws an analogy between a mathematician and a person observing "*a distant range of mountains.*" The key steps in a proof correspond to peaks in the mountain range. To see the proof, one must observe the complete mountain range, i.e., the ridges that link all the peaks. Hardy notes that sometimes the observer

> "*... can distinguish a ridge which vanishes in the distance, and conjectures that it leads to a peak in the clouds or below the horizon.*"

Conjecturing the unknown within a mathematical proof is often referred to as a *eureka step*. The discovery of eureka steps represents one of the major problems for automated theorem proving.

This is particularly true in the case of proof by mathematical induction. Reasoning about recursively defined structures or any form of repetition requires mathematical induction. Inductive proof is therefore crucial for reasoning about the correctness of computer systems. Consequently, techniques for automating inductive reasoning are more than just of academic interest; they have real practical significance to industry.

Inductive proof presents very challenging search control problems for automated reasoning, which give rise to various kinds of eureka steps:

* The research reported in this paper was supported by EPSRC grant GR/J/80702 and ARC grant 438.
** Current address: Department of Computing and Electrical Engineering, Heriot-Watt University, Riccarton, EH14 4AS Edinburgh, Scotland, UK.

- **induction schemata selection:** the search for an inductive proof involves the selection of an appropriate induction schema. The induction schema is instantiated by the given conjecture and an induction variable in the conjecture. All universally quantified variables are candidate induction variables. While the set of possible induction variables is finite, the set of induction schemata is infinite. Consequently, the selection of the induction schema introduces an infinite branching point into the search space.

- **lemma discovery:** the word 'lemma' is used differently in inductive systems from the way it is used in non-inductive systems, e.g., predicate calculus provers. A lemma does not just mean an intermediate result which is generated as a side-effect of search. A lemma may be a separate theorem that is required to complete the original proof. The introduction of such a lemma requires the *cut rule* of inference, i.e.,

$$\frac{\Gamma, \alpha \vdash \beta \quad \Gamma \vdash \alpha}{\Gamma \vdash \beta}$$

  The cut rule can be eliminated from predicate calculus [10] but not from inductive systems. Since the cut rule allows for the introduction of arbitrary new formulae, lemma discovery introduces an infinite branching point into the search space.

- **generalization:** paradoxically, it is sometimes necessary to generalize a conjecture in order for an inductive proof to succeed. Generalization is problematic, however, since it also requires the *cut rule* of inference. Like lemma discovery, generalization therefore introduces an infinite branching point into the search space. In choosing the cut formula we must guard against over-generalization, i.e., attempting to prove a non-theorem.

- **case analyses:** conditional proof is an important technique. Control is necessary, however, to guard against arbitrary case analysis leading to divergence in the search for a proof.

Most inductive theorem provers rely upon user intervention in spotting the need for and in constructing such eureka steps. In contrast, we present a theorem-proving architecture that supports the automatic discovery of eureka steps through the analysis of failed proof attempts. Our approach builds upon *rippling* [7, 4], a heuristic that plays a pivotal role in guiding the search for an inductive proof. We demonstrate how the constraints rippling places on the search space and its declarative nature enable us to automatically patch failed proof attempts through the construction of appropriate eureka steps. In particular, we show how the systematic analysis of the failure of rippling can be used in the selection of induction schemas and the conjecturing of lemmata, generalizations, and case analyses.

1.2. Background

It has been shown how the common structure that defines a family of proofs can be expressed as a *proof plan* [5]. This common structure can be exploited in the search for particular proofs. A proof plan has two complementary components: a proof *method* and a proof *tactic*. By prescribing the structure of a proof at the level of primitive inferences, a tactic [11] provides the guarantee part of the proof. In contrast, a method provides a more declarative explanation of the proof by means of *preconditions*. Each method has associated *effects*. The execution of the effects simulates the application of the corresponding tactic. Theorem proving in the proof planning framework is a two-phase process:

1. Tactic construction is by a process of method composition: Given a goal, an applicable method is selected. The applicability of a method is determined by evaluating the method's preconditions. The method effects are then used to calculate subgoals. This process is applied recursively until no more subgoals remain. Because of the one-to-one correspondence between methods and tactics, the output from this process is a composite tactic tailored to the given goal.
2. Tactic execution generates a proof in the object-level logic. Note that no search is involved in the execution of the tactic. All the search is taken care of during the planning process.

The real benefits of having separate planning and execution phases become apparent when a proof attempt fails. The declarative nature of method preconditions provides a basis for using failure productively. In [6] an extension to the proof planning framework is proposed in which proof *critics* are introduced in order to complement proof methods. The role of the proof critic is to capture patchable exceptions to the proof method. Since a proof method may fail in various ways, each method may be associated with a number of critics. A critic has preconditions and patches. The preconditions of a critic characterise an interesting failure, while the patch prescribes how the failure can be overcome. Critics are able to analyze partial proofs and have a global effect upon the proof process. As well as patching failed proof attempts, critics have also been applied to the problem of identifying and correcting faulty conjectures [17].

1.3. Overview

In this paper we present the use of the critics mechanism [16] in systematically analyzing the failure of the ripple heuristic. In Section 2 a proof plan for induction is outlined in which we emphasize the pivotal role played by rippling. Section 3 forms the core of the paper. It focuses on the ripple heuristic and how its failure can be interpreted productively in the search for inductive proofs. Search control issues relating to the selection and application of patches are discussed in Section 4. A comparison with related techniques is presented in Section 5. The implementation of the proof critics for rippling is outlined in Section 6 together with a discussion of test results. Finally in Sections 7 and 8 we draw our conclusions and outline our plans for future work.

2. A Proof Plan for Induction

In the context of goal-directed proof, the application of a rule of induction generates base- and step-case subgoals. Example induction rules are presented in Figure 1. We will consider rule (3) in detail. The goal-directed application of (3) generates two subgoals, one base-case, i.e.,

$$\mathsf{P}(\mathsf{nil})$$

and one step-case, i.e.,

$$\mathsf{P}(\mathsf{t}) \rightarrow \mathsf{P}(\mathsf{h} :: \mathsf{t})$$

In the step-case $\mathsf{P}(\mathsf{h} :: \mathsf{t})$ is called the *induction conclusion* and $\mathsf{P}(\mathsf{t})$ is called the *induction hypothesis*. The method level structure of our proof plan for induction is given in Figure 2. The role of each method is outlined in the following sections. The heuristic that is embodied within the ripple method underpins the induction proof plan as a whole. That is, while the ripple heuristic controls the rewriting of step-case goals, it also constrains the search for induction rules and the application of induction hypotheses.

$$\frac{\mathsf{P}(0) \quad \forall n : nat.\mathsf{P}(n) \rightarrow \mathsf{P}(s(n))}{\forall x : nat.\mathsf{P}(x)} \tag{1}$$

$$\frac{\mathsf{P}(0) \quad \mathsf{P}(s(0)) \quad \forall n : nat.\mathsf{P}(n) \rightarrow \mathsf{P}(s(s(n)))}{\forall x : nat.\mathsf{P}(x)} \tag{2}$$

$$\frac{\mathsf{P}(nil) \quad \forall h : \tau.\forall t : list(\tau).\mathsf{P}(t) \rightarrow \mathsf{P}(h :: t)}{\forall l : list(\tau).\mathsf{P}(l)} \tag{3}$$

$$\frac{\mathsf{P}(nil) \quad \forall h : \tau.\mathsf{P}(h :: nil) \quad \forall h_1, h_2 : \tau.\forall t : list(\tau).\ \mathsf{P}(t) \rightarrow \mathsf{P}(h_1 :: h_2 :: t)}{\forall l : list(\tau).\ \mathsf{P}(l)} \tag{4}$$

Four example rules of mathematical induction are shown. The first and the third are the structural inductions for natural numbers and lists respectively. Note that s *is the successor function for the natural numbers while* nil *and* :: *are the empty list and list constructor respectively. The second and fourth rules are examples of two step inductions for the naturals and lists respectively.*

Figure 1. Example rules of mathematical induction.

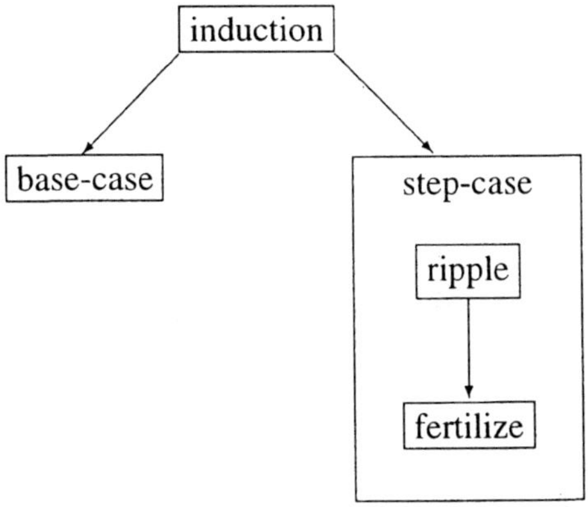

The proof plan for induction consists of the application of a rule of induction followed by one or more base- and step-cases. Only one of each is shown here. In the step-case, ripple controls the rewriting of the induction conclusion so that fertilize can apply the induction hypotheses.

Figure 2. A proof plan for mathematical induction.

The ripple heuristic is based upon the observation that a copy of the induction hypothesis is embedded within the induction conclusion. Following Hutter [15] we call this the *skeleton* term structure. The role of the ripple method is to eliminate the difference between the conclusion and hypothesis while preserving the skeleton term structure. Meta-level annotations are used to express this control information. To illustrate, the annotated version of (3) takes the form

$$\frac{\mathtt{P(nil)} \quad \forall \mathtt{h} : \mathtt{T}.\forall \mathtt{t} : \mathtt{list}(\tau).\ \mathtt{P(t)} \rightarrow \mathtt{P}(\boxed{\mathtt{h} :: \underline{\mathtt{t}}}^{\uparrow})}{\forall \mathtt{l} : \mathtt{list}(\tau).\ \mathtt{P(l)}}$$

The annotated term $\boxed{\mathtt{h} :: \underline{\mathtt{t}}}^{\uparrow}$ is called a *wave*. Note that the box and underlining are meta-level constructions. The underlined object-level term (i.e., $\mathtt{t}$) is called the *wave-hole*. The object-level term structure within the box but excluding the wave-hole (i.e., $\mathtt{h} :: \ldots$) is called the *wave-front*. Wave-fronts highlight the difference between the conclusion and the hypothesis. The arrow is used to indicate the direction of movement of the wave-front within the expression tree of the induction conclusion. The need for directed wave-fronts will be explained in Section 2.3.1. In the following sections we outline the role each component of the proof plan plays in the search for inductive proof.

2.1. Induction Method

As mentioned above, the constraints that the ripple heuristic places upon step-case proof attempts also constrain the choice of induction rule. To illustrate, consider the conjecture

$$\forall t : \mathrm{list}(\tau).\ \mathrm{rev}(\mathrm{rev}(t)) = t \tag{5}$$

where rev denotes list reversal. Note that rev is defined in terms of list concatenation, i.e., $<>$. The definitions of $<>$ and rev give rise to the following rewrite rules:[1]

$$\mathrm{nil} <> Z \Rightarrow Z \tag{6}$$

$$X :: Y <> Z \Rightarrow X :: (Y <> Z) \tag{7}$$

$$\mathrm{rev}(\mathrm{nil}) \Rightarrow \mathrm{nil} \tag{8}$$

$$\mathrm{rev}(X :: Y) \Rightarrow \mathrm{rev}(Y) <> X :: \mathrm{nil} \tag{9}$$

The manipulation of wave-fronts is performed by a syntactic class of rewrite rules called *wave-rules*. Wave-rules are guaranteed to preserve the skeleton term structure while making progress towards applying an induction hypothesis. Rewrite rules (7) and (9) provide a set of wave-rules that include

$$\boxed{X :: \underline{Y}^{\uparrow} <> Z} \Rightarrow \boxed{X :: \underline{(Y <> Z)}}^{\uparrow} \tag{10}$$

$$\mathrm{rev}(\boxed{X :: \underline{Y}}^{\uparrow}) \Rightarrow \boxed{\underline{\mathrm{rev}(Y)} <> X :: \mathrm{nil}}^{\uparrow} \tag{11}$$

The general notion of a wave-rule is explained in Section 2.3.1. In terms of the induction method, wave-rules provide a mechanism for indexing appropriate rules of induction, i.e., induction rules which will enable the application of wave-rules. In the case of (5) induction rule (3) is suggested by the $\mathrm{rev}(\boxed{X :: \underline{Y}}^{\uparrow})$ wave-term appearing on the *LHS* of (11).

2.2. Base-Case Method

The base-case method performs simplification through the use of definitional rewrite rules. For example, a proof of (5) by induction rule (3) generates a base-case subgoal of the form

$$\mathrm{rev}(\mathrm{rev}(\mathrm{nil})) = \mathrm{nil}$$

Two applications of (8) reduces this subgoal to an identity, i.e.,

$$\mathrm{nil} = \mathrm{nil}$$

2.3. Step-Case Method

The ripple and fertilize methods form the step-case method. The role of the ripple method is to rewrite the induction conclusion so that the induction hypothesis can be applied by the fertilize method.[2]

2.3.1. *Ripple Method*

Wave-fronts and wave-rules greatly constrain the search for step-case proofs. The ripple method controls the application of wave-rules. To illustrate, consider the step-case proof obligation associated with the proof of (5). Induction rule (3) gives an induction hypothesis of the form

$$\mathrm{rev}(\mathrm{rev}(t)) = t \tag{12}$$

while the initial induction conclusion takes the form

$$\mathrm{rev}(\mathrm{rev}(\boxed{h :: \underline{t}}^{\uparrow})) = \boxed{h :: \underline{t}}^{\uparrow}$$

Note that the wave-fronts are associated with the induction rule so that the process of annotating an induction conclusion is automatic. The ripple method restricts the rewriting of an induction conclusion to wave-rules. The application of a wave-rule requires that both the object-level and meta-level term structure match. To illustrate, the application of wave-rule (11) to the initial induction conclusion gives rise to

$$\mathrm{rev}(\boxed{\underline{\mathrm{rev}(t)} <> h :: \mathrm{nil}}^{\uparrow}) = \boxed{h :: \underline{t}}^{\uparrow}$$

Wave-rules are not restricted to recursive definitions; for instance,

$$\mathrm{rev}(\boxed{\underline{Y} <> X :: \mathrm{nil}}^{\uparrow}) \Rightarrow \boxed{X :: \underline{\mathrm{rev}(Y)}}^{\uparrow} \tag{13}$$

$$\boxed{X :: \underline{Y}}^{\uparrow} = \boxed{X :: \underline{Z}}^{\uparrow} \Rightarrow Y = Z \tag{14}$$

are wave-rules that are derived from lemmata about rev, $<>$, and list equality. From (13), the induction conclusion becomes

$$\boxed{h :: \underline{\mathrm{rev}(\mathrm{rev}(t))}}^{\uparrow} = \boxed{h :: \underline{t}}^{\uparrow}$$

Finally, by (14) we obtain

$$\mathrm{rev}(\mathrm{rev}(t)) = t \tag{15}$$

All wave-fronts have been eliminated, so the rippling of the conclusion is complete. The conclusion is said to be *fully rippled* and ready to be fertilized.

The strategy as illustrated above is called *longitudinal* rippling. The aim of the strategy is to manipulate the wave-fronts so that they dominate the skeleton term structure of the induction conclusion. Note that the above example is a degenerate case in which the wave-fronts are completely eliminated; that is, they are said to *peter out*.

An alternative strategy is called *transverse* rippling. This strategy exploits the fact that universally quantified variables in the induction hypothesis can be instantiated differently from the corresponding variables in the induction conclusion. To illustrate, consider the following conjecture:

$$\forall t,\ l : list(\tau).\ rev(t) <> l = qrev(t, l) \tag{16}$$

where $qrev$ is the tail recursive version of rev. The definition of $qrev$ provides the following wave-rule:

$$qrev(\boxed{X :: \underline{Y}}^{\uparrow}, Z) \Rightarrow qrev(Y, \boxed{X :: \underline{Z}}^{\downarrow}) \tag{17}$$

Note the change in direction of the wave-fronts. Rippling allows for upward directed wave-fronts to be turned downward but not vice versa. This restriction enables the use of bidirectional rewrite rules without the risk of looping. Other systems, such as Nqthm [2], rely upon the user to indicate which direction a bidirectional rewrite rule should be used. In proving (16), wave-rules (11) and (17) suggest induction on the variable t using rule (3). We concentrate here on the step-case that gives rise to an induction hypothesis of the form[3]

$$rev(t) <> L = qrev(t, L) \tag{18}$$

In order to exploit universally quantified variables, additional meta-level annotations are used. The meta-level construction $\lfloor \ldots \rfloor$ is used to indicate an object-level term within the conclusion, which corresponds to universally quantified variable in the hypothesis, e.g.,

$$rev(\boxed{h :: \underline{t}}^{\uparrow}) <> \lfloor l \rfloor = qrev(\boxed{h :: \underline{t}}^{\uparrow}, \lfloor l \rfloor) \tag{19}$$

These meta-level terms are called *sinks*. The aim of the transverse strategy is to move wave-fronts into sinks. The rewriting of (19), the initial induction conclusion, using wave-rules (11) and (17) gives

$$\boxed{\underline{rev(t)} <> h :: nil}^{\uparrow} <> \lfloor l \rfloor = qrev\left(t, \left\lfloor \boxed{h :: \underline{l}}^{\downarrow} \right\rfloor\right) \tag{20}$$

Now we need to use the associativity of $<>$, which gives rise to a wave-rule of the form

$$\boxed{\underline{X} <> Y}^{\uparrow} <> Z \Rightarrow X <> \boxed{Y <> \underline{Z}}^{\downarrow} \tag{21}$$

The application of (21) to the *LHS* of (20) gives rise to

$$rev(t) <> \left\lfloor \boxed{h :: nil <> \underline{l}}^{\downarrow} \right\rfloor = qrev\left(t, \left\lfloor \boxed{h :: \underline{l}}^{\downarrow} \right\rfloor\right)$$

$$\mathtt{f}(\mathtt{g}(\mathtt{h}(\boxed{\mathtt{c_1}(\underline{\mathtt{x}})}^{\uparrow}),\ \mathtt{k}(\lfloor \mathtt{y} \rfloor)))$$

$$\boxed{\mathtt{c_2}(\underline{\mathtt{f}(\mathtt{g}(\mathtt{h}(\mathtt{x}),\mathtt{k}(\lfloor \boxed{\mathtt{c_3}(\underline{\mathtt{y}})}^{\downarrow} \rfloor))))})}^{\uparrow}$$

A schematic conclusion is shown that gives rise to both longitudinal and transverse ripples.

Figure 3. Rippling: *general pattern.*

Note that both wave-fronts now appear within the scope of sinks. In general, to exploit sinks, the application of a transverse wave-rule may need to be followed by further applications of longitudinal wave-rules. This is called *rippling-in* and typically corresponds to the folding of definitions. Finally, simplification[4] of the wave-front on the *LHS* using rewrite rules (6) and (7) gives

$$\mathtt{rev}(\mathtt{t}) <> \left\lfloor \boxed{\mathtt{h} :: \underline{\mathtt{l}}}^{\downarrow} \right\rfloor = \mathtt{qrev}\left(\mathtt{t},\ \left\lfloor \boxed{\mathtt{h} :: \underline{\mathtt{l}}}^{\downarrow} \right\rfloor\right) \tag{22}$$

The rippling of the induction conclusion is now complete. The general pattern of the longitudinal and transverse strategies is presented in Figure 3. A detailed description of rippling appears in [4].

2.3.2. *Fertilize Method*

The fertilize method controls the application of induction hypotheses. In the case of conjecture (5) fertilization is trivial, since the hypothesis (12) and the fully rippled conclusion (15) are identical. At most, matching is required, as illustrated in conjecture (16) where the matching of the hypothesis (18) against the fully rippled conclusion (22) instantiates $\mathtt{L}$ to be $\mathtt{h} :: \mathtt{l}$.

3. Productive Use of Failure

We now consider how an inductive proof might fail. In particular we consider failure of the ripple method. To do this, however, we must consider the ripple method in more detail. The actual application of wave-rules is controlled by the *wave* method. The ripple method iterates over the wave method. The preconditions for the application of a longitudinal wave-rule are presented in Figure 4, while those for applying transverse wave-rules are presented in Figure 5.

In the following four sections we systematically analyze the ways in which preconditions of the wave method can fail, and we present the patches suggested in each case.

Preconditions:
1. $\mathsf{wave\_term(Conc, Poc, LHS)}$
2. $\mathsf{wave\_rule\_match(Rn, long(D), Cond \rightarrow LHS \Rightarrow RHS, Subs)}$
3. $\mathsf{tautology(Hyps \vdash Cond)}$

Definition of meta-logical terms:
- Hyps and Conc denote the current hypotheses and conclusion respectively;
- $\mathsf{wave\_term}(\mathsf{T}, \mathsf{P}, W)$ means that W is the wave-term at position P within a term T;
- $\mathsf{wave\_rule\_match(N, T(D), C \rightarrow L \Rightarrow R, S)}$ means that N is the name of a wave-rule of type T with rewrite direction D, which unifies with the term L. S is the set of substitutions for any higher-order meta variables instantiated by the unification.
- $\mathsf{tautology(S)}$ is true when the sequent S is a tautology.

Figure 4. Wave method: *longitudinal-rippling.*

Preconditions:
1. $\mathsf{wave\_term(Conc, Pos, LHS)}$
2. $\mathsf{wave\_rule\_match(Rn, trans(D), Cond \rightarrow LHS \Rightarrow RHS, Subs)}$
3. $\mathsf{tautology(Hyps \vdash Cond)}$
4. $\mathsf{sinkable(RHS)}$

Definition of meta-logical terms:
- $\mathsf{sinkable(T)}$ term T contains a wave-front that is directed towards a sink.
- all other terms are as defined in Figure 4.

Figure 5. Wave method: *transverse-rippling.*

3.1. INDUCTION REVISION CRITIC

As mentioned in Section 2.1, rippling can be used in the selection of induction rules. We now examine how the ripple heuristic can select an inappropriate induction rule. Consider the conjecture

$$\forall \mathsf{t}, \mathsf{l} : \mathsf{list}(\tau).\ \mathsf{even}(\mathsf{length}(\mathsf{t} <> \mathsf{l})) \leftrightarrow \mathsf{even}(\mathsf{length}(\mathsf{l} <> \mathsf{t})) \tag{23}$$

where the predicate *even* holds for the even natural numbers and *length* computes the length of a list. The definitions of *even* and *length* provide wave-rules that include

$$\mathsf{even}(\boxed{\mathsf{s}(\mathsf{s}(\underline{\mathsf{X}}))}^{\uparrow}) \Rightarrow \mathsf{even}(\mathsf{X}) \tag{24}$$

$$\mathsf{length}(\boxed{\mathsf{X} :: \underline{\mathsf{Y}}}^{\uparrow}) \Rightarrow \boxed{\mathsf{s}(\underline{\mathsf{length}(\mathsf{Y})})}^{\uparrow} \tag{25}$$

We assume an additional lemma that prodives a wave-rule of the form

$$\mathrm{length}(X <> \boxed{Y :: \underline{Z}}^{\uparrow}) \Rightarrow \boxed{s(\underline{\mathrm{length}(X <> Z)})}^{\uparrow} \tag{26}$$

In proving (23), wave-rule (10) suggests[5] an induction on t using rule (3). The initial induction conclusion takes the form

$$\mathrm{even}(\mathrm{length}(\boxed{h :: \underline{t}}^{\uparrow} <> l)) \leftrightarrow \mathrm{even}(\mathrm{length}(l <> \boxed{h :: \underline{t}}^{\uparrow}))$$

Using wave-rules (10), (25), and (26) the conclusion rewrites to give

$$\underbrace{\mathrm{even}(\boxed{s(\underline{\mathrm{length}(t <> l)})}^{\uparrow})}_{\text{blocked}} \leftrightarrow \underbrace{\mathrm{even}(\boxed{s(\underline{\mathrm{length}(l <> t)})}^{\uparrow})}_{\text{blocked}} \tag{27}$$

No more wave-rules are applicable, so the wave method fails to apply. This corresponds to the failure of precondition 2 of the wave method. However, the wave-fronts are preventing the fertilize method from applying. To patch this failure, we could attempt to introduce additional wave-fronts by refining our chosen induction rule. For this to be profitable, there must exist a wave-rule that partially matches with one of the blocked wave-terms, i.e., a wave-rule with an *LHS* of the following form:

$$\mathrm{even}(\boxed{s(\boxed{F_1(\underline{\mathrm{length}(Y <> \boxed{F_2(\underline{Z})}^{\uparrow})})}^{\uparrow})}^{\uparrow}) \Rightarrow \cdots \tag{28}$$

where F_1 and F_2 are second-order meta variables. Note that we use dotted boxes the denote *potential* wave-fronts, i.e., wave-fronts that may or may not exist. The *LHS* of wave-rule (24) unifies with the *LHS* of (28), instantiating F_1 and F_2 to be $\lambda x.s(x)$ and $\lambda x.x$, respectively. Discussion of higher-order unification is delayed until Section 4.3.

This success suggests the need for an additional wave-front in overcoming the blocked goal. This analysis is expressed as a critic in Figure 6. The application of the critic to goal (27) suggests that an additional wave-front of the form $\boxed{s(\underline{\ldots})}^{\uparrow}$ is required. The rippling-in of the composite wave-front, i.e., $\boxed{s(s(\underline{\ldots}))}^{\downarrow}$, using wave-rules (25) and (10), suggests a two-step induction, i.e., the initial selection of rule (3) is replaced by rule (4). This gives rise to a revised induction conclusion of the form

$$\mathrm{even}(\mathrm{length}(\boxed{h_1 :: h_2 :: \underline{t}}^{\uparrow} <> l)) \leftrightarrow \mathrm{even}(\mathrm{length}(l <> \boxed{h_1 :: h_2 :: \underline{t}}^{\uparrow}))$$

This revision enables the induction conclusion to be fully-rippled, i.e.,

$$\mathrm{even}(\mathrm{length}(\boxed{h_1 :: \underline{(\boxed{h_2 :: \underline{t}}^{\uparrow} <> l)}}^{\uparrow})) \leftrightarrow \mathrm{even}(\boxed{s(\underline{\mathrm{length}(l <> \boxed{h_2 :: \underline{t}}^{\uparrow})})}^{\uparrow})$$

Preconditions:

1. Precondition 1 of wave method holds, i.e.,
   $\mathit{wave\_term}(\_, \mathrm{Pos}, \mathit{WaveTerm})$
2. Preconditions 2 and 3 of wave method fail, i.e.,
   no matching wave-rule
   no condition to check
3. $\mathtt{partial\_wave\_rule\_match}(\mathit{WaveTerm}, \mathit{RevisedWaveTerm})$

Patch:

1. $\mathtt{revise\_induction}(\mathit{RevisedWaveTerm}, \mathrm{Pos}, \mathrm{Conc}, \mathrm{IndRule})$
2. $\mathtt{propagate\_induction}(\mathrm{Plan}, \mathrm{IndRule})$

Definition of meta-logical terms:

- Conc and Plan denote the current conclusion and plan structure, respectively;
- $\mathtt{partial\_wave\_rule\_match}(T, W)$ holds when W is the wave-front suggested by a partial wave-rule match with the term T;
- $\mathtt{revise\_induction}(W, P, G, R)$ R is the induction rule suggested by the wave-term W at position P within the goal G;
- $\mathtt{propagate\_induction}(P, R)$ propagates R, the revised induction suggestion through the plan structure P.

Figure 6. Wave critic: *induction revision.*

$$\mathrm{even}(\boxed{\mathrm{s}(\underline{\mathrm{length}(\boxed{\mathrm{h}_2 :: \underline{\mathrm{t}}}^{\uparrow} <> \mathrm{l})})}^{\uparrow}) \leftrightarrow \mathrm{even}(\boxed{\mathrm{s}(\mathrm{s}(\underline{\mathrm{length}(\mathrm{l} <> \mathrm{t})}))}^{\uparrow})$$

$$\mathrm{even}(\boxed{\mathrm{s}(\mathrm{s}(\underline{\mathrm{length}(\mathrm{t} <> \mathrm{l})}))}^{\uparrow}) \leftrightarrow \mathrm{even}(\mathrm{length}(\mathrm{l} <> \mathrm{t}))$$

$$\mathrm{even}(\mathrm{length}(\mathrm{t} <> \mathrm{l})) \leftrightarrow \mathrm{even}(\mathrm{length}(\mathrm{l} <> \mathrm{t}))$$

The exception to the general pattern of rippling, which leads to induction revision, is captured in Figure 7.

3.2. Lemma Discovery Critic

The proof of conjecture (23), presented in Section 3.1, relied upon a lemma that provided wave-rule (26). In this section we focus upon the problem of discovering such lemmata automatically. Consider the following term:

$$\underbrace{\ldots \mathrm{rev}(\boxed{\underline{\mathrm{rev}(\mathrm{t})} <> \mathrm{h} :: \mathrm{nil}}^{\uparrow})}_{\mathrm{blocked}} \ldots \tag{29}$$

The rippling of this term blocked if we assume that wave-rule (13) is not available. This corresponds to the failure of precondition 2 of the wave method.

$$f(g(h(\boxed{c_1(\underline{x})}^{\uparrow}), k(\lfloor y \rfloor))) \qquad f(g(h(\boxed{c_1(c_2(\underline{x}))}^{\uparrow}), k(\lfloor y \rfloor)))$$

$$\underbrace{f(\boxed{c_3(\underline{g(h(x), \boxed{c_4(\underline{k(\lfloor y \rfloor)})}^{\downarrow})})}^{\uparrow})}_{\text{missing wave-fronts}} \qquad f(\boxed{c_3(c_5(\underline{g(h(x), \boxed{c_6(c_4(\underline{k(\lfloor y \rfloor)}))}^{\downarrow})}))}^{\uparrow})$$

$$\boxed{c_7(\underline{f(g(h(x), k(\lfloor \boxed{c_8(\underline{y})}^{\downarrow} \rfloor))))})}^{\uparrow}$$

before after

Conclusion schemata are shown before and after the application of the induction revision critic. The rippling of the initial conclusion becomes blocked. Partial wave-rule matching identifies missing wave-fronts, i.e., $\boxed{c_3(c_5(\underline{\ldots}))}^{\uparrow}$ *and* $\boxed{c_6(c_4(\underline{\ldots}))}^{\downarrow}$, *which are used to calculate a revised induction suggestion, i.e.,* $\boxed{c_1(c_2(\underline{x}))}^{\uparrow}$.

Figure 7. Exception: *missing wave-fronts.*

The induction revision critic, through precondition 2, will initiate a search for a wave-rule of the form

$$rev(\boxed{\underline{F_1(\underline{rev(Y)})}^{\uparrow} <> X :: nil}^{\uparrow}) \Rightarrow \cdots$$

Assuming only the wave-rules that are provided by the definition of rev, then the search will fail. This failure means that any revision of the current induction will lead to a potentially infinite divergence in the proof, i.e.,

$$\ldots rev(\boxed{(\underline{rev(t)} <> h_2 :: nil) <> h_1 :: nil}^{\uparrow})\ldots$$

$$\ldots rev(\boxed{((\underline{rev(t)} <> h_3 :: nil) <> h_2 :: nil) <> h_1 :: nil}^{\uparrow})\ldots$$

$$\ldots rev(\boxed{(((\underline{rev(t)} <> h_4 :: nil) <> h_3 :: nil) <> h_2 :: nil) <> h_1 :: nil}^{\uparrow})\ldots$$

$$\vdots \qquad\qquad \vdots \qquad\qquad \vdots$$

Note that Nqthm will diverge in this way. The failure of the partial wave-rule match (Precondition 3, Figure 6) suggests that we should instead be looking for a missing wave-rule, i.e., a lemma. This analysis is expressed as a critic in Figure 8.

Preconditions:

1. Precondition 1 of wave method holds, i.e., $wave\_term(\_, \_, WaveTerm)$
2. Preconditions 2 and 3 of wave method fail, i.e., *no matching wave-rule* *no condition to check*
3. $not\ partial\_wave\_rule\_match(WaveTerm, \_)$

Patch:

1. $construct\_lemma\_lhs(WaveTerm, Sequent, LHS)$
2. $construct\_lemma\_rhs(WaveTerm, Sequent, RHS)$
3. $generalize\_lemma(LHS \Rightarrow RHS, Rewrite)$
4. $validate\_lemma(Rewrite)$

Definition of meta-logical terms:

- $Sequent$ is the current goal sequent;
- $construct\_lemma\_lhs(W, S, L)$ L is the left-hand-side of a wave-rule based upon the wave-term W and the current goal sequent S;
- $construct\_lemma\_rhs(W, S, R)$ R is the right-hand-side of a wave-rule based upon the wave-term W and the current goal sequent S;
- $generalize\_lemma(R, G)$ G is a generalization of the rewrite rule R;
- $validate\_lemma(R)$ records schematic wave-rules derived from R and verifies instantiations generated for R during subsequent proof planning.

Figure 8. Wave critic: *lemma discovery (speculation).*

The patch associated with the critic involves the search for a lemma and its proof. We view this as a four-step process, which, in outline, involves

1. construction of the left-hand-side of the lemma,
2. construction of the right-hand-side of the lemma,
3. generalization of the conjectured lemma, and
4. proof of the generalized lemma.

This process may involve backtracking. For example, step 3 may be resatisfied if an over-generalization is detected by the failure of step 4.

We have two strategies for implementing this process. The first is called *lemma calculation* and is the less general of the two strategies but involves little search. The second strategy, *lemma speculation*, is very general and as a consequence introduces additional search control problems. These alternative strategies are presented in the following two sections.

3.2.1. *Lemma Calculation*

Lemma calculation is applicable when we are in a position to partially apply an induction hypothesis. To illustrate, consider again the step-case proof of conjecture (5) presented in Section 2.3.1. The effect of not having wave-rule (13) means that the *LHS* of the conclusion becomes blocked, i.e.,

$$\underbrace{rev(\boxed{\underline{rev(t)} <> h :: nil}^{\uparrow})}_{\text{blocked}} = \boxed{h :: \underline{t}}^{\uparrow} \tag{30}$$

Fortunately the *RHS* is fully rippled. That is, the wave-hole on the *RHS* contains a copy of the *RHS* of (12), the induction hypothesis. Consequently, the induction hypothesis can be used to rewrite the conclusion. We now consider the construction of a wave-rule to unblock the *LHS* of the goal based upon this kind of partial use of the induction hypothesis:

1. The blocked wave-term defines the *LHS* of the missing wave-rule. In the case of goal (30) there is one blocked wave-term, which gives an *LHS* of the form

$$rev(\boxed{\underline{rev(t)} <> h :: nil}^{\uparrow}) \Rightarrow \cdots$$

2. The *RHS* of the missing wave-rule is constructed by rewriting the fully rippled subterm of the blocked conclusion using the induction hypothesis. In the current example this means rewriting the wave-hole associated with $\boxed{h :: \underline{t}}^{\uparrow}$ using (12), which gives

$$\cdots \Rightarrow \boxed{h :: \underline{rev(rev(t))}}^{\uparrow}$$

3. The generalization of the wave-rule resulting from steps 1 and 2 takes the form

$$rev(\boxed{\underline{Y} <> X :: nil}^{\uparrow}) \Rightarrow \boxed{X :: \underline{rev(Y)}}^{\uparrow}$$

   Generalization is desirable to maximize the applicability of wave-rules and to simplify the proof of the underlying lemma. Further discussion of generalization is delayed until Section 4.1.2.

4. The proof of the underlying lemma can be performed either before or after the wave-rule is applied in the main proof. In practice it is performed before.

In general lemmata may be conditional. Currently we use the context in which a ripple proof becomes blocked to suggest appropriate conditions.

3.2.2. *Lemma Speculation*

Lemma speculation is required where calculation is not applicable. For example, consider the conjecture

$$\forall t, l : list(\tau).\ rev(rev(t) <> l) = rev(l) <> t \tag{31}$$

In proving (31), wave-rule (11) suggests[6] induction on t using rule (3). The initial induction conclusion takes the form

$$rev(rev(\boxed{h :: \underline{t}}^{\uparrow}) <> \lfloor l \rfloor) = rev(\lfloor l \rfloor) <> \boxed{h :: \underline{t}}^{\uparrow}$$

By wave-rule (11) this rewrites to

$$\underbrace{rev\,(\boxed{\underline{rev(t)} <> h :: nil}^{\uparrow} <> \lfloor l \rfloor)}_{\text{blocked}} = \underbrace{rev(\lfloor l \rfloor) <> \boxed{h :: \underline{t}}^{\uparrow}}_{\text{blocked}} \tag{32}$$

Lemma calculation is not applicable because neither of the wave-fronts is fully rippled. Now we consider the construction of a wave-rule to unblock the *RHS* of (32). Note that we could equally have selected the *LHS*:

1. The *LHS* of the missing wave-rule is constructed as for lemma calculation, i.e.,

$$rev(l) <> \boxed{h :: \underline{t}}^{\uparrow} \Rightarrow \cdots$$

2. From the skeleton-preserving property of rippling we know that the *RHS* of the missing wave-rule must fit the following general form

$$\cdots \Rightarrow \cdots (\cdots rev(l) \cdots <> t) \cdots$$

   We exploit the skeleton-preserving property to construct a schematic *RHS*; that is, second-order meta variables are used to specify the missing wave-front structure. Transverse, longitudinal, and hybrid forms must be considered:

$$\cdots \Rightarrow \boxed{F_1(\underline{rev(l)}, h, t, l)}^{\downarrow} <> t$$

$$\cdots \Rightarrow \boxed{G_1(\underline{rev(l) <> t}, h, t, l)}^{\uparrow}$$

$$\cdots \Rightarrow \boxed{G_1(\underline{\boxed{F_1(\underline{rev(l)}, h, t, l)}^{\downarrow} <> t}, h, t, l)}^{\uparrow}$$

3. The generalization of the schematic wave-rules resulting from steps 1 and 2 is

$$W <> \boxed{X :: \underline{Y}}^{\uparrow} \Rightarrow \boxed{F_1(\underline{W}, X, Y)}^{\downarrow} <> Y \tag{33}$$

$$W <> \boxed{X :: \underline{Y}}^{\uparrow} \Rightarrow \boxed{G_1(\underline{W <> Y}, X, Y)}^{\uparrow} \tag{34}$$

$$W <> \boxed{X :: \underline{Y}}^{\uparrow} \Rightarrow \boxed{G_1(\underline{\boxed{F_1(\underline{W}, X, Y)}^{\downarrow} <> Y}, X, Y)}^{\uparrow} \tag{35}$$

   An application of a schematic wave-rule and its instantiation go hand in hand. Rippling constrains the process of instantiation; that is, any instantiations for meta variables must preserve the skeleton term structure. With the addition of schematic wave-rules (33), (34), and (35), the wave method is now applicable to (32). Using wave-rule (33) the goal becomes

$$\underbrace{\mathrm{rev}\left(\boxed{\underline{\mathrm{rev}(t)} <> h :: nil}^{\uparrow} <> \lfloor l \rfloor\right)}_{\text{blocked}} = \boxed{F_1(\underline{\mathrm{rev}(\lfloor l \rfloor)}, h, t)}^{\downarrow} <> t$$

The directionality of the potential wave-front and the associated wave-hole constrains the search for an applicable wave-rule. In this case wave-rule (11) can be applied in reverse, i.e.,

$$\boxed{\underline{\mathrm{rev}(Y)} <> X :: nil}^{\downarrow} \Rightarrow \mathrm{rev}\left(\boxed{X :: \underline{Y}}^{\downarrow}\right)$$

The *RHS* of the induction conclusion becomes

$$\cdots = \boxed{F_3(\underline{\mathrm{rev}(\lfloor \boxed{F_2(\mathrm{rev}(l), h, t) :: \underline{l}}^{\downarrow} \rfloor)}, h, t)}^{\downarrow} <> t \tag{36}$$

The associated unification instantiates F_1 to be

$$\lambda w.\lambda x.\lambda y.F_3(w <> F_2(w, x, y) :: nil, x, y)$$

Wave-rule schema (33) now becomes

$$W <> \boxed{X :: \underline{Y}}^{\uparrow} \Rightarrow \boxed{F_3(\underline{\boxed{\underline{W} <> F_2(W, X, Y) :: nil}^{\downarrow}}, X, Y)}^{\downarrow} <> Y$$

Each incremental instantiation of a wave-rule schema is coupled with an attempt to coerce the remaining meta variables. This coercion process involves exploring possible projections. Where a meta variable denotes a potential wave-front, only the projection of the wave-hole is considered. A conjecture disprover is employed to filter the candidates generated by the coercion process. The disprover is conservative; it may fail to identify invalid instantiations, but it will not reject results that are in fact correct. However, all instantiated wave-rule (lemma) schemas are submitted to the prover for verification. In the current example, coercion is successful and instantiates F_2 and F_3 to be $\lambda w.\lambda x.\lambda y.x$ and $\lambda w.\lambda x.\lambda y.w$, respectively. This completes the instantiation of wave-rule schema (33). The resulting wave-rule takes the form

$$W <> \boxed{X :: \underline{Y}}^{\uparrow} \Rightarrow \boxed{\underline{W} <> X :: nil}^{\downarrow} <> Y \tag{37}$$

Note that the process of constructing (37) completes the rippling of the *RHS* of the conclusion

$$\underbrace{\mathrm{rev}\left(\boxed{\underline{\mathrm{rev}(t)} <> h :: nil}^{\uparrow} <> \lfloor l \rfloor\right)}_{\text{blocked}} = \mathrm{rev}\left(\left\lfloor \boxed{h :: \underline{l}}^{\downarrow} \right\rfloor\right) <> t$$

If we use (37) in reverse, i.e.,

$$\underbrace{\begin{array}{c} f(g(h(\boxed{c_1(\underline{x})}^{\uparrow}),\ k(\lfloor y \rfloor))) \\ f(g(\boxed{c_2(\underline{h(x)})}^{\uparrow},\ k(\lfloor y \rfloor))) \end{array}}_{\text{missing wave-rule}} \qquad \begin{array}{c} f(g(h(\boxed{c_1(\underline{x})}^{\uparrow}),\ k(\lfloor y \rfloor))) \\ f(g(\boxed{c_2(\underline{h(x)})}^{\uparrow},\ k(\lfloor y \rfloor))) \end{array}$$

$$f(\boxed{F_1(\underline{g(h(x), \boxed{F_2(\underline{k(\lfloor y \rfloor)}, x, y)}^{\downarrow}}), x, y)}^{\uparrow})$$

before after

Conclusion schemata are shown before and after the application of the lemma discovery (speculation) critic. The rippling of the initial conclusion becomes blocked by the term $g(\ldots, k(\lfloor y \rfloor))$. *The patch is to construct a schematic wave-rule, which will unblock the ripple. Further rippling will then instantiate the schema.*

Figure 9. Exception: *missing wave-rule.*

$$\boxed{\underline{W} <> X :: nil}^{\uparrow} <> Y \Rightarrow W <> \boxed{X :: \underline{Y}}^{\downarrow}$$

the rippling of the *LHS* of the conclusion can also be completed:

$$rev(rev(t) <> \lfloor \boxed{h :: \underline{l}}^{\downarrow} \rfloor) = rev(\lfloor \boxed{h :: \underline{l}}^{\downarrow} \rfloor) <> t$$

4. The proof of the underlying lemma has to be delayed until the wave-rule schema is fully instantiated.

The exception to the general pattern of rippling, which leads to the search for new wave-rules, is presented in Figure 9.

3.3. Generalization Critic

It is sometimes the case that a conjecture must be generalized before we can prove it by induction; that is, the generalization leads to a stronger induction hypothesis, which allows a proof to go through. Generalization through the introduction of accumulator variables is an example of such a phenomenon. To illustrate, consider the following specialization of conjecture (16):

$$\forall t : list(\tau).\ rev(t) = qrev(t, nil)$$

Wave-rules (11) and (17) suggest an induction on t using rule (3). In the step-case we get an induction hypothesis of the form

$$rev(t) = qrev(t, nil) \tag{38}$$

while the initial induction conclusion takes the form

$$\mathtt{rev}(\boxed{h :: \underline{t}}^{\uparrow}) = \mathtt{qrev}(\boxed{h :: \underline{t}}^{\uparrow}, \mathtt{nil})$$

Using wave-rule (11) the conclusion rewrites to give

$$\boxed{\underline{\mathtt{rev}(t)} <> h :: \mathtt{nil}}^{\uparrow} = \underbrace{\mathtt{qrev}(\boxed{h :: \underline{t}}^{\uparrow}, \mathtt{nil})}_{\mathtt{blocked}} \tag{39}$$

Note that while wave-rule (17) matches the wave-term on the *RHS*, its applicability is ruled out because of the absence of a sink. In terms of the wave method this corresponds to the failure of Precondition 4 (see Figure 5). This failure is captured as a critic in Figure 10. The patch associated with this critic is the generalization of the goal through the introduction of an accumulator variable into the original conjecture. This is achieved with the use of second-order meta variables. The revised conjecture takes the form

$$\forall t,\ l : \mathtt{list}(\tau).F_1(\mathtt{rev}(t), l) = \mathtt{qrev}(t, G_1(l)) \tag{40}$$

Note that the relationship between the accumulator variable l and the original term structure of the conjecture is partially specified using the meta variables F_1 and G_1. The position of $G_1(l)$ within the *RHS* is determined by the position of the wave-front on the *RHS* of wave-rule (17). The insertion of the meta variables

---

Preconditions:

1. Precondition 1 of wave method holds, i.e.,
   $\mathtt{wave\_term}(\_, Pos, WaveTerm)$
2. Precondition 2 of wave method holds, i.e.,
   $\mathtt{wave\_rule\_match}(Rn, \mathtt{trans}(\_), Cond \rightarrow LHS \Rightarrow RHS, \_)$
3. Precondition 3 of wave method holds, i.e.,
   $\mathtt{tautology}(Hyps \vdash Cond)$
4. Precondition 4 of wave method fails, i.e.,
   no sink.

Patch:

1. $\mathtt{generalize\_goal}(Conc, Pos, Rn, GenConc)$
2. $\mathtt{propagate\_generalization}(Plan, GenConc)$

Definition of meta-logical terms:

- $\mathtt{generalize\_goal}(C, P, R, G)$ constructs G a generalization of C based upon the transverse wave-rule R and the position P within C where it is applicable;
- $\mathtt{propagate\_generalization}(P, G)$ propagates the generalized goal G through the plan structure P.

Figure 10. Wave critic: *generalization.*

is automatic. With the revised conjecture an induction on t is again proposed. The induction hypotheses becomes

$$F_1(rev(t), L) = qrev(t, G_1(L)) \tag{41}$$

Note that (41) is stronger than (38), the original hypothesis. The induction conclusion associated with the new induction hypothesis takes the form

$$F_1(rev(\boxed{h :: \underline{t}}^{\uparrow}), \lfloor l \rfloor) = qrev(\boxed{h :: \underline{t}}^{\uparrow}, G_1(\lfloor l \rfloor))$$

As before, wave-rule (11) is applicable to the *LHS* and refines the conclusion as follows:

$$F_1(\boxed{\underline{rev(t)} <> h :: nil}^{\uparrow}, \lfloor l \rfloor) = qrev(\boxed{h :: \underline{t}}^{\uparrow}, G_1(\lfloor l \rfloor))$$

The presence of the sink on the *RHS* means that wave-rule (17) is now applicable, giving

$$F_1(\boxed{\underline{rev(t)} <> h :: nil}^{\uparrow}, \lfloor l \rfloor) = qrev(t, \boxed{h :: \underline{G_1(\lfloor l \rfloor)}}^{\downarrow}) \tag{42}$$

From wave-rule (21) the *LHS* can be further rippled to give

$$rev(t) <> (\boxed{h :: nil <> \underline{F_2(rev(t), \lfloor l \rfloor)}}^{\downarrow}) = qrev(t, \boxed{h :: \underline{G_1(\lfloor l \rfloor)}}^{\downarrow})$$

The effect of this wave-rule application is to instantiate F_1 to be

$$\lambda x.\lambda y.x <> F_2(x, y)$$

As described in Section 3.2.2, after each incremental instantiation of a meta variable we attempt to coerce the remaining meta variables through projection. Here we have the added constraint that sinks corresponding to the same variable must be instantiated consistently. In this example the coercion process instantiates F_2 and G_1 to be $\lambda x.\lambda y.y$ and $\lambda x.x$, respectively. This completes the rippling of the induction conclusion:

$$rev(t) <> (\lfloor \boxed{h :: \underline{l}}^{\downarrow} \rfloor) = qrev(t, \lfloor \boxed{h :: \underline{l}}^{\downarrow} \rfloor)$$

Note the instantiation of (40) corresponds to (16). The exception to the general pattern of rippling caused by missing sinks is presented in Figure 11.

3.4. Case Analyses Critic

The last critic deals with the failure of Precondition 3 of the wave method, i.e., where the condition attached to a wave-rule is not provable in the current context. To illustrate, consider the conjecture

$$\forall a : \tau.\forall t,\ l : list(\tau).\ a \in t \rightarrow a \in (t <> l) \tag{43}$$

$$f(g(h(\boxed{c(\underline{x})}^{\uparrow}), \ldots)) \qquad f(g(h(\boxed{c(\underline{x})}^{\uparrow}), F(\lfloor y \rfloor)))$$

$$\underbrace{f(g(\boxed{c_1(\underline{h(x)})}^{\uparrow}, \ldots))}_{\text{missing sink}} \qquad f(g(\boxed{c_1(\underline{h(x)})}^{\uparrow}, F(\lfloor y \rfloor)))$$

$$f(g(h(x), \boxed{c_2(\underline{F(\lfloor y \rfloor)})}^{\downarrow}))$$

before after

Conclusion schemata are shown before and after the application of the generalization critic. The rippling of the initial conclusion becomes blocked because of the absence of a sink. A meta variable F *is used to introduce a new object-level variable* y. *The positioning within the goal structure of the annotated term* $F(\lfloor y \rfloor)$ *is determined by the RHS of an applicable transverse wave-rule.*

Figure 11. Exception: *missing sink.*

where $\in$ denotes list membership. The definition of $\in$ provides the following rewrite rules:

$$X \in nil \Rightarrow false$$

$$X \neq Y \rightarrow X \in \boxed{Y :: \underline{Z}}^{\uparrow} \Rightarrow X \in Z \tag{44}$$

$$X = Y \rightarrow X \in \boxed{Y :: \underline{Z}}^{\uparrow} \Rightarrow true \tag{45}$$

Note that (45) is not a wave-rule because it is not skeleton preserving. It is instead classified as a *complementary* rewrite rule, since its condition forms part of a covering set that includes the condition associated with (44).

In proving (43), wave-rule (44) suggests an induction on t using rule (3). This provides an induction hypothesis of the form

$$a \in t \rightarrow a \in (t <> L)$$

and an induction conclusion of the form

$$a \in \boxed{h :: \underline{t}}^{\uparrow} \rightarrow a \in (\boxed{h :: \underline{t}}^{\uparrow} <> \lfloor l \rfloor)$$

By wave-rule (10) this rewrites to give

$$\underbrace{a \in \boxed{h :: \underline{t}}^{\uparrow}}_{\text{blocked}} \rightarrow \underbrace{a \in \boxed{h :: \underline{t <> \lfloor l \rfloor}}^{\uparrow}}_{\text{blocked}}$$

Although (44) matches both wave-terms, the associated condition, $a \neq h$, is not provable in the current context. This corresponds to the failure of Precondition 3

Preconditions:

1. Precondition 1 of wave method holds, i.e.,
   $\mathtt{wave\_term}(\_, \_, \mathtt{LHS})$
2. Precondition 2 of wave method holds, i.e.,
   $\mathtt{wave\_rule\_match}(\mathtt{Rn}, \_, \mathtt{Cond} \rightarrow \mathtt{LHS} \Rightarrow \_, \_)$
3. Precondition 3 of wave method fails, i.e.,
   Cond not provable given Hyps

Patch:

1. $\mathtt{setof}(\mathtt{Cond}, (\mathtt{wave\_rule\_match}(\_, \_, \mathtt{Cond} \rightarrow \mathtt{LHS} \Rightarrow \_, \_) \vee$
   $\mathtt{comp\_rule\_match}(\_, \mathtt{Cond} \rightarrow \mathtt{LHS} \Rightarrow \_)),$
   $\mathtt{Conds})$
2. $\mathtt{case\_split}(\mathtt{Plan}, \mathtt{Conds})$

Definition of meta-logical terms:

- $\mathtt{comp\_rule\_match}(\mathtt{N}, \mathtt{R})$ means that N is the name of the complementary rewrite rule R;
- $\mathtt{case\_split}(\mathtt{P}, \mathtt{C})$ introduces a case-split on C at the current node in the plan structure P.

Figure 12. Wave critic: *missing condition.*

of the wave method. The critic associated with this failure is given in Figure 12. The patch is to perform a case analysis based upon the covering set of conditions defined the wave- and complementary-rules. In the current example this covering set gives rise to a case split on $(a \neq h) \vee (a = h)$. By (44), in the $a \neq h$ case, we get

$$a \in t \rightarrow a \in (t <> \lfloor l \rfloor)$$

to which fertilize is applicable. In the $a = h$ case, using (45), we get

$$\mathtt{true} \rightarrow \mathtt{true}$$

which is trivially true. The exception to the general pattern of rippling caused by a missing condition is presented in Figure 13.

3.5. Summary

We have argued that by having an explicit proof plan one is able to accurately pinpoint and interpret failures productively. In the case of the proof plan for induction, the breakdown of the preconditions for rippling provide a handle on the major eureka steps associated with inductive proof. The relationship between precondition failures and patches is summarized in Table I.

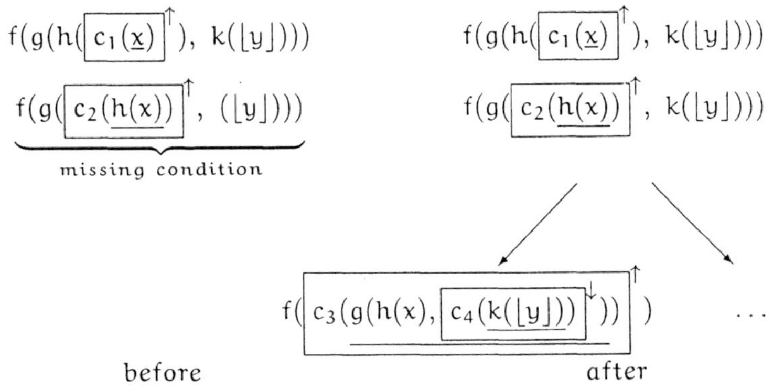

Conclusion schemata are shown before and after the application of the case analysis critic. The rippling of the initial conclusion becomes blocked by the absence of a condition. A set of conditional wave-rules and complementary rewrite rules suggests the missing case split.

Figure 13. Exception: *missing condition.*

Table I. Precondition failures and patches for rippling

Precondition	Generalization	Case Analysis	Induction Revision	Lemma Discovery
1	⋆	⋆	⋆	⋆
2	⋆	⋆	∘	•
3	⋆	•		
4	•			

The association between precondition failure and patches for the ripple heuristic are shown. Note that ⋆, ∘, and • denote success, partial success, and failure, respectively.

4. Search Control Issues

In this section we address a number of search control issues that arise in the critics mechanism and the use of higher-order meta variables in the search for proofs.

4.1. Wave-Term Selection

For a given goal the selection of wave-terms is determined by the *LHS* of an applicable wave-rule. In the absence of an applicable wave-rule we require a mechanism for selecting wave-terms. The search space associated with this selection process is discussed below.

4.1.1. *Most Nested Wave-Fronts*

A blocked goal may contain many blocked wave-fronts. In searching for a wave-front to unblock, we select a *most nested* wave-front. The motivation is that unblocking a more nested wave-front may in turn unblock a less-nested wave-front. To illustrate, consider the goal

$$\ldots \mathit{even}(\boxed{s(\underline{x + \boxed{s(\underline{x})}^{\uparrow}})}^{\uparrow})\ldots$$

Both wave-fronts are blocked if we assume only the wave-rules provided by the definitions of *even* and +, i.e., wave-rules (24) and

$$\boxed{s(\underline{X})}^{\uparrow} + Y \Rightarrow \boxed{s(\underline{X + Y})}^{\uparrow}$$

If our search for a missing wave-rule is successful, i.e.,

$$X + \boxed{s(\underline{Y})}^{\uparrow} \Rightarrow \boxed{s(\underline{X + Y})}^{\uparrow}$$

then the unblocking of the most nested wave-front also enables the unblocking of the less nested wave-front, i.e.,

$$\ldots \mathit{even}(\boxed{s(\underline{x + \boxed{s(\underline{x})}^{\uparrow}})}^{\uparrow})\ldots$$

$$\ldots \mathit{even}(\boxed{s(s(\underline{x + x}))}^{\uparrow})\ldots$$

$$\ldots \mathit{even}(x + x)\ldots$$

4.1.2. *Wave-Term Context*

The context of the wave-term is important when searching for a missing lemma. On the one hand, we wish to minimize the context associated with a wave-rule; on the other hand, we are interested only in wave-rules derived from theorems. For example, consider the following blocked goal:

$$\ldots \mathit{even}(\mathit{length}(l <> \boxed{h :: \underline{t}}^{\uparrow}))\ldots$$

Selecting the minimal wave-term, i.e.,

$$l <> \boxed{h :: \underline{t}}^{\uparrow}$$

will lead to the following invalid wave-rule

$$W <> \boxed{X :: \underline{Y}}^{\uparrow} \Rightarrow \boxed{X :: \underline{W <> Y}}^{\uparrow}$$

The required wave-term takes the form

$$\mathtt{length}(l <> \boxed{h :: \underline{t}}^{\uparrow})$$

and provides a valid wave-rule of the form

$$\mathtt{length}(W <> \boxed{X :: \underline{Y}}^{\uparrow}) \Rightarrow \boxed{s(\underline{\mathtt{length}(W <> Y)})}^{\uparrow}$$

Backtracking is therefore necessary to allow for the selection of progressively larger subterms of the skeleton term structure of the goal. In addition, we also consider the need to manipulate multiple wave-fronts simultaneously. Since the goal term structure is finite, the associated search space is also finite.

Finally, generalization is an important technique in minimizing context. We exploit two forms of generalization: *replacement of common terms* and *distinguishing variables apart*. In the context of automatic inductive theorem proving Aubin [1] was the first to investigate such techniques. We adopt similar heuristics to those developed by Aubin, e.g., exploiting *primary recursion paths*[7] is distinguishing between variables to generalization apart.

All of the above heuristic may lead to invalid wave-rule suggestions. In practice we rely upon a conjecture disprover to prune the search space of wave-rule suggestions.

4.2. Schematic Wave-Rule Construction

As observed in Section 3.2.2 there may be a choice as to whether to apply longitudinal, transverse, or hybrid wave-rule schemata. The strategy we employ is to pursue the direction that provides greatest constraints in the instantiation of higher-order meta variables. For example, in the case of goal (32), where three schematic wave-rule are generated, schema (32) was selected because it gave rise to further rippling.

In general, however, it may not be obvious which is the best schema to apply. We therefore allow backtracking over the selection process using a conjecture disprover to prune the search space.

4.3. Guiding Unification

As illustrated by the lemma discovery and generalization critics, the ability to apply wave-rules in the context of higher-order meta variables is crucial to our technique. The instantiation of such meta variables occurs as a side effect of applying wave-rules. The generality and controllability of such a technique is dependent upon the type of variables used. If we choose first-order variables, then control is not a problem, but the framework is not very general [15]. Alternatively, using higher-order variables we gain generality but at the cost of controllability.

Meta-level control information provides a basis for regaining controllability. Hesketh [13] used the meta-level control information embedded within the wave-front annotations in controlling the search for generalizations. We adopt a similar aproach to Hesketh in using the wave-fronts and wave-holes to divide the higher-order unification task into a number of subtasks. We extend the approach, however, by exploiting the directionality of wave-fronts in focusing the unification process. That is, when rippling-in we match wave-holes first and when rippling-out we first match the superterm that contains the wave-front. This is necessary in order to support the incremental instantiation of meta variables.

Backtracking over alternative instantiations, however, is necessary. This is particularly the case for the generalization critic where we employ an iterative deepening planner. To illustrate the problem, consider again the *LHS* of goal (42), i.e.,

$$F_1(\boxed{\underline{\mathrm{rev}(t)} <> h :: \mathrm{nil}}^{\uparrow}, \lfloor l \rfloor) = \cdots \tag{46}$$

In Section 3.3 the associativity of $<>$ (wave-rule (21)) was used to obtain

$$\mathrm{rev}(t) <> (\boxed{h :: \mathrm{nil} <> \underline{F_2(\mathrm{rev}(t), \lfloor l \rfloor)}}^{\downarrow}) = \cdots$$

However, the associativity of $<>$ provides another applicable wave-rule, i.e.,

$$X <> \boxed{\underline{Y} <> Z}^{\uparrow} \Rightarrow \boxed{\underline{X <> Y} <> Z}^{\uparrow}$$

which, when unified with the *LHS* of (46), gives

$$\boxed{\underline{F_2(\mathrm{rev}(t), \lfloor l \rfloor) <> \mathrm{rev}(t)} <> h :: \mathrm{nil}}^{\uparrow} = \cdots$$

This longitudinal ripple is rejected because it gives rise to a non-theorem that is detected by our conjecture disprover. Control is hard for generalization, since we are dealing with a schematic skeleton term structure. Consequently, the skeleton preservation property of rippling is less useful for restricting search than in the case of lemma discovery.

Higher-order annotated unification is a hard problem. We acknowledge that our approach is far from ideal. The interested reader is referred to [14] for some promising results we may be able to exploit in our future work.

4.4. Critic Selection

Multiple critics may be applicable to a particular proof failure. Preference is given to the critic with the most restrictive preconditions (see Table I). For example, the generalization critic was used to patch goal (39) although lemma calculation was also applicable. However, patches are not guaranteed to succeed so backtracking across critics is supported.

5. Comparison with Related Techniques

Most theorem provers rely upon user interaction to identify, interpret, and patch failures. In contrast, proof critics attempt to automate this process. In order to draw a comparison with other related techniques, we must consider the individual critics presented here.

First, induction revision appears to be unique in theorem-proving terms. In relation to other inductive theorem-proving systems, this kind of patching involves the user in supplying a hint in the form of a dummy recursive function that reflects the desired induction.

Second, lemma calculation is based upon the strategy of cross fertilization [2] and weak-fertilization [4]. A practical limitation of both these strategies is that they discover lemmas in-line. As a consequence the same lemma may be re-discovered and verified many times during the course of a proof. Lemma calculation factors out lemmas and verifies them as separate proof attempts. This eliminates the redundancy problem mentioned above and provides a degree of modularity in structuring proofs.

Third, lemma speculation was first presented in [16]. A missing lemma typically causes divergence in the rewriting of step-case proofs. Lemma speculation preempts such divergence. We are able to achieve this because of the constraints rippling imposes on the search space. In contrast, Thomas and Jantke [20] observe divergence patterns and use them to suggest generalizations in the context of inductive completion. A similar idea has been applied by Walsh [21] to the problem of proof divergence in the SPIKE [3] theorem prover. Divergence patterns generated by SPIKE are used to suggest lemmas. Central to this approach is the technique of *difference matching* [9]. Difference matching is used to speculate the structure of missing lemmas by generalizing over a sequence of diverging formulae. This process of overcoming divergence is largely independent of the SPIKE proof strategy. In contrast, the proof critics mechanism has direct access to the meta-level control information which is used to guide the search for proofs. This increases the sophistication by which failure can be interpreted. For example, without our meta-level notion of a 'sink' it is hard to see how one could distinguish between the need for generalization as opposed to a missing lemma. The diversity of patches associated with the ripple heuristic demonstrates the power of rippling and the proof critics mechanism. More generally the proof critics mechanism supports the global analysis of the proof process, since it has access to the whole proof structure (see Section 5 of [16]). As a consequence, we believe that the approach presented here provides a powerful framework for patching proofs.

Fourth, the use of higher-order meta variables in discovering generalizations through the introduction of accumulators was first achieved by Hesketh [13]. The generalization critic presented here represents a rational reconstruction and extension of this work.

Lastly, the case analyses critic is an alternative to having a method for supporting conditional rewriting [4]. The critic mechanism, however, allows for more sophisticated case analyses involving the composition of multiple partial proofs. This idea is illustrated in [16].

6. Implementation and Results

The proof critics presented here have been implemented and tested. Our implementation is an extension of the CLAM [6] proof planning system and exploits the higher-order features of λ-Prolog [18].

Our test results are presented in Tables II, III, and IV. The proofs of all the example conjectures[8] given in Table II are discovered completely automatically. These proofs are based only upon definitions supplied by the user. Except for the generalization examples, all additional lemmas are automatically discovered by the system.

The example conjectures in Table II are classified under the four critics. In the case of lemma discovery, conjectures T1 to T13, T22 to T26, and T48 to T50 required only the relatively weak strategy of lemma calculation. Even so, Nqthm failed[9] to prove T1, T3, T4, T6, T7, T8, T9, T10, T11, T13, and T22 through to T26. Examples T14 to T21 required lemma speculation while T27 to T35 required generalization. Note that different generalizations were obtained depending upon the available lemmata. All the examples that required induction revision, lemma speculation, or generalization fall into a class of inductive theorem that are not uncommon but are acutely difficult to prove. With the exception of T14, Nqthm failed to prove any of these examples. Finally, examples T22 to T26 and T48 to T50 illustrate the need for multiple critics in patching some conjectures.

7. Limitations and Further Work

We have focussed upon failure in the context of constructor style induction. The complementary destructor style induction is already incorporated within rippling. A destructor style induction introduces wave-fronts into the induction hypothesis, e.g.,

$$\mathsf{G}[\mathsf{f}(\boxed{\mathsf{d}(\underline{\mathsf{x}})}^{+}, \mathsf{y})] \vdash \mathsf{G}[\mathsf{f}(\mathsf{x}, \mathsf{y})]$$

Rippling uses *creational* wave-rules [4] to set-up a ripple in the conclusion by *neutralizing* the wave-fronts in the hypothesis, e.g.,

$$\mathsf{G}[\mathsf{f}(\boxed{\mathsf{d}(\underline{\mathsf{x}})}^{+}, \mathsf{y})] \vdash \mathsf{G}[\boxed{\mathsf{c}(\underline{\mathsf{f}(\boxed{\mathsf{d}(\underline{\mathsf{x}})}^{-}, \mathsf{y})})}^{\uparrow}]$$

Creational wave-rules have the following general form:

$$Cond \rightarrow \mathsf{f}(\mathsf{x}, \mathsf{y}) \Rightarrow \boxed{\mathsf{c}(\underline{\mathsf{f}(\boxed{\mathsf{d}(\underline{\mathsf{x}})}^{-}, \mathsf{y})})}^{\uparrow}$$

Table II. Example conjectures

No.	Conjecture	(i)	(ii)	(iii)	(iv)
T1	double(X)=X+X		L1		
T2	length(X<>Y)=length(Y<>X)		L2		
T3	length(X<>Y)=length(Y)+length(X)		L1		
T4	length(X<>X)=double(length(X))		L2		
T5	length(rev(X))=length(X)		L3		
T6	length(rev(X<>Y))=length(X)+length(Y)		L2		
T7	length(qrev(X,Y))=length(X)+length(Y)		L1		
T8	nth(X,nth(Y,Z))=nth(Y,nth(X,Z))		L4&L5		
T9	nth(W,nth(X,nth(Y,Z)))=nth(Y,nth(X,nth(W,Z)))		L6&L7		
T10	rev(rev(X))=X		L8		
T11	rev(rev(X)<>rev(Y))=Y<>X		L9&10		
T12	qrev(X,Y)=rev(X)<>(Y)		L11		
T13	half(X+X)=X		L1		
T14	ordered(isort(X))		L12		
T15	X+s(X)=s(X+X)		L1		
T16	even(X+X)		L1		
T17	rev(rev(X<>Y))=rev(rev(X))<>rev(rev(Y))		L8		
T18	rev(rev(X)<>Y)=rev(Y)<>X		L11&L13		
T19	rev(rev(X))<>Y=rev(rev(X<>Y))		L8		
T20	even(length(X<>X))		L2		
T21	rotate(length(X),X<>Y)=Y<>X		L11&L13		
T22	even(length(X<>Y))↔even(length(Y<>X))	(3)↦(4)	L14		
T23	half(length(X<>Y))=half(length(Y<>X))	(3)↦(4)	L15		
T24	even(X+Y)↔even(Y+X)	(1)↦(2)	L16		
T25	even(length(X<>Y))↔even(length(Y)+length(X))	(3)↦(4)	L16		
T26	half(X+Y)=half(Y+X)	(1)↦(2)	L17		
T27	rev(X)=qrev(X,nil)			G1	
T28	revflat(X)=qrevflat(X,nil)			G2	
T29	rev(qrev(X,nil))=X			G3&G4	
T30	rev(rev(X)<>nil)=X			G5&G6	
T31	qrev(qrev(X,nil),nil)=X			G7&G8	
T32	rotate(length(X),X)=X			G9	
T33	fac(X)=qfac(X,1)			G10	
T34	X∗Y=mult(X,Y,0)			G11	
T35	exp(X,Y)=qexp(X,Y,1)			G12	
T36	X∈Y→X∈(Y<>Z)				⋆
T37	X∈Z→X∈(Y<>Z)				⋆
T38	(X∈Y)∨(X∈Z)→X∈(Y<>Z)				⋆
T39	X∈nth(Y,Z)→X∈Z				⋆
T40	X⊂Y→(X∪Y=Y)				⋆
T41	X⊂Y→(X∩Y=X)				⋆
T42	X∈Y→X∈(Y∪Z)				⋆

Table II. Continued

No.	Conjecture	(i)	(ii)	(iii)	(iv)
T43	X∈Y→X∈(Z∪Y)				★
T44	(X∈Y)∧(X∈Z)→(X∈Y∩Z)				★
T45	X∈insert(X,Y)				★
T46	X=Y→(X∈insert(Y,Z)↔true)				★
T47	X≠Y→(X∈insert(Y,Z)↔X∈Z)				★
T48	length(isort(X))=length(X)		L18		★
T49	X∈isort(Y)→X∈Y		L19		★
T50	count(X,isort(Y))=count(X,Y)		L20&L21		★

The numbered columns denote (i) *induction revision,* (ii) *lemma discovery,* (iii) *generalization and* (iv) *case split. Note that* nth(X,Y) *denotes the list constructed by removing the first* Xth *elements from* Y. *Note also that* fac, exp *and* x *denote factorial, exponentiation and multiplication while* qfac, qexp *and* mult *denote tail recursive versions, respectively.*

Table III. Lemmata

No.	Lemma
L1	X+s(Y)=s(X+Y)
L2	length(X<>Y::Z)=s(length(X<>Z))
L3	length(X<>Y::nil)=s(length(X))
L4	nth(s(W),nth(X,Y::Z))=nth(W,nth(X,Z))
L5	nth(s(V),nth(s(W),X::Y::Z))=nth(s(V),nth(W,X::Z))
L6	nth(s(V),nth(W,nth(X,Y::Z)))=nth(V,nth(W,nth(X,Z)))
L7	nth(s(U),nth(V,nth(s(W),X::Y::Z)))=nth(s(U),nth(V,nth(W,X::Z)))
L8	rev(X<>(Y::nil))=Y::rev(X)
L9	rev(X<>(Y<>Z::nil))=Z::rev(X<>Y)
L10	rev((X<>Y::nil)<>nil)=Y::rev(X<>nil)
L11	(X<>(Y::nil))<>Z=X<>(Y::Z)
L12	ordered(Y)→ordered(insert(X,Y))
L13	(X<>Y)<>Z::nil=X<>(Y<>Z::nil)
L14	even(length(W<>Z))↔even(length(W<>X::Y::Z))
L15	length(W<>X::Y::Z)=s(s(length(W<>Z)))
L16	even(X+Y)↔even(X+s(s(Y)))
L17	X+s(s(Y))=s(s(X+Y))
L18	length(insert(X,Y)=s(length(Y))
L19	(X≠Y→(X∈insert(Y,Z)→X∈Z)
L20	count(X,insert(X,Y))=s(count(X,Y))
L21	X≠Y→(count(X,insert(Y,Z))=count(X,Z))
L22	(X<>Y)<>Z=X<>(Y<>Z)
L23	(X*Y)*Z=X*(Y*Z)
L24	(X+Y)+Z=X+(Y+Z)

We are currently investigating the automatic discovery of creational wave-rules. This will involve generalizing our technique for generating conditional lemmata.

Table IV. Generalized conjectures

No.	Generalization	Lemmata
G1	rev(X)<>Y=qrev(X,Y)	L22
G2	revflat(X)<>Y=qrevflat(X,Y)	L22
G3	rev(qrev(X,Y))=rev(Y)<>X	L11
G4	rev(qrev(X,rev(Y)))=Y<>X	L8&L11
G5	rev(rev(X)<>Y)=rev(Y)<>X	L11
G6	rev(rev(X)<>rev(Y))=Y<>X	L8&L11
G7	qrev(qrev(X,Y),nil)=rev(Y)<>X	L11
G8	qrev(qrev(X,rev(Y)),nil)=Y<>X	L8&L11
G9	rotate(length(X),X<>Y)=Y<>X	L11&L22
G10	fac(X)*Y=qfac(X,Y)	L23
G11	(X*Y)+Z=mult(X,Y,Z)	L24
G12	exp(X,Y)*Z=qexp(X,Y,Z)	L23

The lemmata used to motivate each generalization are indicated in the right-hand column.

There is a strong connection between creational wave-rules and well-founded inductions. The discovery of missing creational wave-rules may provide a handle on the problem of generating new inductions dynamically.

In terms of the proof critics mechanism, the analysis presented here relies upon patterns of precondition failures that are constructed off-line. We are currently investigating the use of meta-level inference in deriving patches automatically. This will provide a more dynamic form of failure analysis.

8. Conclusions

The ripple method has proved to be a very successful search control heuristic within inductive proof. We have shown, through the use of proof critics, that rippling provides useful heuristic guidance even when a proof attempt fails.

More generally, it has been argued [8] that the separation of meta-level control information from the object-level logic brings clarity, flexibility, and modularity to reasoning systems as well as providing a more constrained search space. The proof critics technique demonstrates another advantage of this separation, namely, *robustness*, that is, the ability to exploit failure productively in the search for proofs.

Notes

[1] We use $\Rightarrow$ to denote rewrite rules and $\rightarrow$ to denote logical implication.

[2] The phrase *fertilization*, which describes the use of hypotheses, was introduces in [2].

[3] Note that we replace l in the hypothesis with L. We use upper-case letters to denote variables, while constants are denoted by lower-case letters.

[4] The simplification of wave-fronts is guaranteed to preserve the skeleton term structure.

[5] Actually t or l are equally good candidate induction variables. The analysis presented here works for either variable.

[6] Note that an induction on ι would be equally suitable and that the technique being presented works for either suggestion.

[7] We exploit *ripple paths*, a generalization of the recursion path notion.

[8] The examples come from a number of sources that include [1, 2, 19], and [21].

[9] This is without the aid of the linear arithmetic decision procedure. With the decision procedure, Nqthm is able to prove T1, T3, T6, and T7. More generally, when we talk of 'failure' we are talking about the failure of Nqthm to find a proof without user intervention, i.e., additional lemmata, induction hints, or generalizations supplied by the user.

Acknowledgements

We are indebted to David Basin for his encouragement and detailed comments during the preparation of this paper. We also thank Ian Frank, Erica Melis, Raul Monroy, Julian Richardson, Toby Walsh, and three anonymous JARS referees for feedback on this paper.

References

1. Aubin, R.: Some generalization heuristics in proofs by induction, in G. Huet and G. Kahn (eds), *Actes du Colloque Construction: Amélioration et vérification de Programmes*, Institut de recherche d'informatique et d'automatique, 1975.
2. Boyer, R. S. and Moore, J. S.: *A Computational Logic*, Academic Press, New York, 1979. ACM monograph series.
3. Bouhoula, A. and Rusinowitch, M.: Automatic case analysis in proof by induction, in *Proc. 13th IJCAI, Int. Joint Conf. Artificial Intelligence*, 1993.
4. Bundy, A., Stevens, A., van Harmelen, F., Ireland, A., and Smaill, A.: Rippling: A heuristic for guiding inductive proofs, *Artificial Intelligence* **62** (1993), 185–253. Also available from Edinburgh as DAI Research Paper No. 567.
5. Bundy, A.: The use of explicit plans to guide inductive proofs, in R. Lusk and R. Overbeek (eds), *9th Conf. Automated Deduction*, Springer-Verlag, 1988, pp. 111–120. Longer version avilable from Edinburgh as DAI Research Paper No. 349.
6. Bundy, A., van Harmelen, F., Horn, C., and Smaill, A.: The Oyster-Clam system, in M. E. Stickel (ed.), *10th Int. Conf. Automated Deduction*, Lecture Notes in Artificial Intelligence 449, Springer-Verlag, 1990, pp. 647–648. Also available from Edinburgh as DAI Research Paper No. 507.
7. Bundy, A., van Harmelen, F., Smaill, A., and Ireland, A.: Extensions to the rippling-out tactic for guiding inductive proofs, in M. E. Stickel (ed.), *10th Int. Conf. Automated Deduction*, Lecture Notes in Artificial Intelligence 449, Springer-Verlag, 1990, pp. 132–146. Also available from Edinburgh as DAI Research Paper No. 459.
8. Bundy, A. and Welham, B.: Using meta-level inference for selective application of multiple rewrite rules in algebraic manipulation, *Artificial Intelligence* **16**(2) (1981), 189–212. Also available from Edinburgh as DAI Research Paper No. 121.
9. Basin, D. and Walsh, T.: Difference matching, in Deepak Kapur (ed.), *11th Conf. Automated Deduction*, Saratoga Springs, NY, USA, June 1992. Published as Springer Lecture Notes in Artificial Intelligence 607, pp. 295–309.
10. Gentzen, G.: *The Collected Papers of Gerhard Gentzen*, edited by M. E. Szabo, North-Holland, Amsterdam, 1969.
11. Gordon, M. J., Milner, A. J., and Wadsworth, C. P.: *Edinburgh LCF – A Mechanised Logic of Computation*, Lecture Notes in Computer Science 78, Springer-Verlag, 1979.
12. Hardy, G. H.: Mathematical proof, *MIND: A Quarterly Review of Psychology and Philosophy*, **38**(149) (January 1929), 1–25.
13. Hesketh, J. T.: Using middle-out reasoning to guide inductive theorem proving, PhD thesis, University of Edinburgh, 1991.

14. Hutter, D. and Kohlhase, M.: A Colored Version of the λ-Calculus, Seki-report sr-95-05, University of Saarland, 1995.
15. Hutter, D.: Guiding inductive proofs, in M. E. Stickel (ed.), *10th Int. Conf. Automated Deduction*, Lecture Notes in Artificial Intelligence 449, Springer-Verlag, 1990, pp. 147–161.
16. Ireland, A.: The use of planning critics in mechanizing inductive proofs, in A. Voronkov (ed.), *Int. Conf. Logic Programming and Automated Reasoning, LPAR 92, St. Petersburg*, Lecture Notes in Artificial Intelligence 624, Springer-Verlag, 1992, pp. 178–189. Also available from Edinburgh as DAI Research Paper No. 592.
17. Monroy, R., Bundy, A., and Ireland, A.: Proof plans for the correction of false conjectures, in F. Pfenning (ed.), *5th Int. Conf. Logic Programming and Automated Reasoning, LPAR 94*, Lecture Notes in Artificial Intelligence 822, Springer-Verlag, 1994, pp. 54–68. Also available from Edinburgh as Research Paper No. 681.
18. Miller, D. and Nadathur, G.: An overview of λProlog, in R. Bowen and K. Kowalski (eds), *Proc. 5th Int. Logic Programming Conf./5th Symp. on Logic Programming*, MIT Press, 1988.
19. Manna, Z. and Waldinger, R.: *The Logical Basis for Computer Programming*, Vol. 1. *Deductive Reasoning*, Addison/Wesley, Reading, MA, 1985.
20. Thomas, M. and Jantke, K. P.: Inductive inference for solving divergance in Knuth–Bendix, in K. P. Jantke (ed.), *Analogical and Inductive Inference, Proc. AII'89*, Springer-Verlag, 1989, pp. 288–303.
21. Walsh, T.: A divergence critic, in Alan Bundy (ed.), *12th Conf. Automated Deduction*, Lecture Notes in Artificial Intelligence 814, Springer-Verlag, Nancy, France, 1994.

Journal of Automated Reasoning **16:** 113–145, 1996.

Middle-Out Reasoning for Synthesis and Induction

INA KRAAN*
Institut für Informatik, Universität Zürich, CH-8057 Zürich, Switzerland
e-mail: inak@ifi.unizh.ch

DAVID BASIN**
Max-Planck-Institut für Informatik, Saarbrücken, Germany
e-mail: basin@mpi.mpg.de

ALAN BUNDY‡
Department of Artificial Intelligence, University of Edinburgh, Scotland, U.K.
e-mail: bundy@ai.ed.ac.uk

Abstract. We develop two applications of middle-out reasoning in inductive proofs: logic program synthesis and the selection of induction schemes. Middle-out reasoning as part of proof planning was first suggested by Bundy et al. Middle-out reasoning uses variables to represent unknown terms and formulae. Unification instantiates the variables in the subsequent planning, while proof planning provides the necessary search control.

Middle-out reasoning is used for synthesis by planning the verification of an unknown logic program: The program body is represented with a meta-variable. The planning results both in an instantiation of the program body and a plan for the verification of that program. If the plan executes successfully, the synthesized program is partially correct and complete.

Middle-out reasoning is also used to select induction schemes. Finding an appropriate induction scheme during synthesis is difficult because the recursion of the program, which is unknown at the outset, determines the induction in the proof. In middle-out induction, we set up a schematic step case by representing the constructors that are applied to induction variables with meta-variables. Once the step case is complete, the instantiated variables correspond to an induction appropriate to the recursion of the program.

We have implemented these techniques as an extension of the proof planning system *CLAM*, called *Periwinkle*, and synthesized a variety of programs fully automatically.

Key words: Automated theorem proving, proof planning, induction, logic program synthesis, meta-variables, higher-order unification.

* Supported by the Swiss National Science Foundation and ARC Project BC/DAAD Grant 438. The work described in this paper was carried out while the first author was at the Department of Artificial Intelligence of the University of Edinburgh.

** Supported by the German Ministry for Research and Technology (BMFT) under grant ITS 9102 and ARC Project BC/DAAD Grant 438. Responsibility for the contents of this publication lies with the authors.

‡ Supported by SERC grant GR/J/80702, ESPRIT BRP grant 6810, ESPRIT BRP grant EC-US 019-76094, and ARC Project BC/DAAD Grant 438.

1. Introduction

We develop techniques based on *proof planning* and *middle-out reasoning* that enable the automatic synthesis of logic programs. Proof planning entails explicit reasoning about how to construct proofs. Middle-out reasoning allows proof planning to progress even though an object being reasoned about is not yet fully known. Middle-out reasoning represents unspecified objects in the proof with variables and instantiates them using unification. Thus we can plan proofs while leaving certain unknown terms or formulae to be filled in at a later stage. In program synthesis based on inductive proofs, two things are unknown: first, most obviously, the program to be synthesized; second, the type of induction used in the proof. This is because the appropriate type of induction depends on the type of recursion of the program to be synthesized.

Middle-out reasoning for program synthesis and induction has been implemented as an extension of the proof planning system *CIAM* [16]. The extended system, called *Periwinkle*, has been used to synthesize a variety of programs. *Periwinkle* is available on request from the first author.

This paper elaborates and extends [30, 31]; more detail can be found in [29]. The paper is organized as follows: Section 2 is an introduction to proof planning. Sections 3 and 4 present middle-out reasoning for logic program synthesis and the selection of induction schemes. Section 5 presents new methods that proved necessary for synthesis. Section 6 reports on the implementation and on practical results. Section 7 presents ideas for further research, and Section 8 draws conclusions.

2. Proof Planning

To use built-in heuristics common in theorem provers more flexibly, Bundy [10] suggests using a meta-logic to reason about and to plan proofs. Proof plans are constructed in the meta-logic by successively applying *methods* to a conjecture until a combination of methods has been found that forms a complete plan. A method is a partial specification of a tactic [23] in the following sense: If a sequent matches the input pattern, and the preconditions are met, the tactic is applicable; if the tactic succeeds, the output conditions will be true of the resulting sequents. Explicit proof planning has been implemented in *CIAM* [16], which constructs plans for inductive proofs in a variant Martin-Löf type theory [36]. The plans are executable in *Oyster* [16], a sequent-style interactive proof checker.

The advantages of the meta-logic approach are that the search for proofs takes place at the meta-level rather than the object level. The search is less expensive, since methods capture the effects of the corresponding tactics, while avoiding the possibly considerable cost of executing them. More important, however, the meta-level representation of the proof can be augmented with additional information on the proof to restrict the search space. The information is passed from method to method, which gives a global rather than a local view of the proof.

Proof planning has concentrated on inductive proofs. The central method for inductive proofs is `ind_strat`, a composite method capturing the structure of such proofs. It is composed of the methods `induction`, `base_case`, and `step_case`. The `induction` method selects a set of induction variables and an induction scheme for a given conjecture, a step crucial to the success of proof planning. The `induction` method uses *recursion analysis* to select an induction. Recursion analysis is a rational reconstruction and extension of the heuristics used in NQTHM to select induction variables and schemes [9, 41, 15]. Recursion analysis prefers induction variables that occur in the recursive positions of the function or relation dominating them (i.e., are smaller in the recursive calls) and which can be rewritten using an axiom or a lemma. It selects a scheme that corresponds to the recursion of the dominating function or relation. In essence, recursion analysis is a look-ahead into the rewriting of the step case.

The `induction` method applies recursion analysis to the input sequent. It succeeds if the analysis suggests a suitable induction scheme and fails otherwise. For the associativity of $+$, for example,

$$\forall x, y, z.(x + y) + z = x + (y + z) \tag{1}$$

recursion analysis using

$$\forall x, y.s(x) + y = s(x + y) \tag{2}$$

$$\forall x, y.x = y \rightarrow s(x) = s(y) \tag{3}$$

suggests structural induction on x. Outputs are the base and step cases for the selected induction. Step cases are annotated for the `ripple` method (see below).

The `base_case` method iterates over a symbolic evaluation method `sym_eval` and a simplification and tautology-checking method `elementary`.

In the step case, the main objective is to rewrite the induction conclusion so that the induction hypothesis can be exploited. The `step_case` method applies the `ripple` method to rewrite the induction conclusion and the `fertilize` method to exploit the induction hypothesis. The `ripple` method embodies the *rippling* heuristic [14]. This heuristic uses rewrite rules to eliminate the differences between the induction hypothesis and the induction conclusion so that the induction hypothesis can be exploited. The function symbols that appear in the conclusion, but not in the hypothesis, are called *wave fronts*. Initially, the wave fronts immediately dominate the induction variables. The role of rippling is to move them outwards – just like ripples on a lake – until a perfect reflection of the induction hypothesis is left. The rippling heuristic has been shown to terminate [14, 3]. We represent wave fronts as boxes with holes. The holes are indicated by underlinings. For the step case of the proof of (1), the `induction` method sets up the annotated sequent

$$(x + y) + z = x + (y + z) \vdash \left(\boxed{s(\underline{x})} + y\right) + z = \boxed{s(\underline{x})} + (y + z).$$

If we remove the structure in the non-underlined parts of the boxes from the conclusion, we obtain the *skeleton*, i.e., a copy of the induction hypothesis.

Rippling consists of applying annotated rewrite rules called *wave rules*. The annotations on wave rules ensure that applying a wave rule will move at least one wave front up in the term tree of the induction conclusion if the annotations in the rule are compatible with those of the conclusion. The (simplified) schematic format of a wave rule that moves one wave front is

$$F\left[\boxed{S[\underline{U}]}\right] :\Rightarrow \boxed{T[\underline{F[U]}]}.$$

The effect of applying a wave rule is to move the wave front S on the left-hand side outwards past the F and to turn it into a wave front T on the right-hand side, whose position is higher up the term tree. Note that $:\Rightarrow$ indicates rewriting, not implication. In inductive theorem proving, rippling reasons backwards from the induction conclusion to the induction hypothesis. Thus, rewrite rules may be based on equality, equivalence, or implication from right to left. Wave rules based on (2) and (3), for example, are

$$\boxed{s(\underline{M})} + N :\Rightarrow \boxed{s(\underline{M + N})}$$

$$\boxed{s(\underline{M})} = \boxed{s(\underline{N})} :\Rightarrow M = N,$$

where M and N are free variables.

Rippling as presented so far is known as *rippling out*. It is an extension of the *ripple-out* heuristic developed by Aubin [1]. For a complete description of all variations of rippling see [14].

The fertilization methods exploit the induction hypothesis. If, after rippling, the wave front surrounds the entire induction conclusion (or has disappeared), the `strong fertilization` method appeals to the induction hypothesis directly. If the wave fronts do not yet surround the entire induction conclusion, the `weak fertilization` method uses the induction hypothesis as a rewrite rule.

Rippling may terminate before the induction hypothesis can be exploited. We then say that the rippling is blocked. There are various techniques to *unblock* the rippling, which modify the conclusion in some way that makes a wave rule or fertilization applicable. A common unblocking step is simplifying a wave front.

3. Middle-Out Synthesis

3.1. Pure Logic Programs

The logic programs we synthesize are the completions of a subset of normal programs (see Lloyd [34]), which we call *pure logic programs*. They are similar to pure logic programs as defined by Bundy et al. in [13] and the logic descriptions of Deville [19]. Formally, we define them as finite sets of pure logic program clauses. A *pure logic program clause* is a closed, typed formula of the form

$$\forall \overline{x}\colon \overline{T}.\, A(\overline{x}) \leftrightarrow H,$$

where $\overline{x}$ is a vector of distinct variables with types given by $\overline{T}$ (generally left implicit in the following), $A(\overline{x})$ is an atom, called the head of a clause, and H is a *Horn body*. A formula H is a Horn body if, in Backus-Naur notation,

$$H ::= A \mid H_1 \wedge H_2 \mid H_1 \vee H_2 \mid \exists x.\, H,$$

where A is an atom whose name is a known relation (such as $=$ or $\neq$) or whose name is among the names of the heads of previously defined clauses (including the one being defined). An example of a pure logic program is

$$\forall x, l.\, member(x, l) \leftrightarrow \exists h, t.\, l = h :: t \wedge (x = h \vee member(x, t)) \quad (4)$$

$$\forall i, j.\, subset(i, j) \leftrightarrow i = nil \vee \exists h, t.\, i = h :: t \wedge member(h, j) \wedge subset(t, j). \quad (5)$$

We synthesize pure logic programs because they are a suitable intermediate representation between non-executable specifications and executable programs. In particular, the definition of pure logic programs guarantees that

$$\forall \overline{x} : \overline{T}.\, A(\overline{x}) \leftarrow H$$

corresponds to a set of definite program clauses (see Lloyd [34]).

The class of pure logic programs is very general: It captures the semantics of pure logic programming languages. It also captures the basic recursive structure of algorithms, while avoiding non-logical aspects such as order of execution and non-logical primitives, which are normally specific to the implementation of a logic programming language. This enables us to break down the formidable task of synthesis: First, we synthesize the basic structure of the algorithm, independent of any particular programming language. In a second step, we can translate the pure logic program into a logic programming language of our choice and introduce non-logical primitives as desired. Synthesizing pure logic programs has another advantage: The intended meaning of the program coincides with its logical meaning. Thus, we can reason within the well-understood framework of (many-sorted) first-order logic (with induction) and bring knowledge in theorem-proving to bear.

Using pure logic programs, we can prove partial correctness and completeness[1] by showing that $A(\overline{x})$ as defined by $A(\overline{x}) \leftrightarrow H$ is equivalent to S

$$\forall \overline{x} : \overline{T}.\, A(\overline{x}) \leftrightarrow H \vdash \forall \overline{x} : \overline{T}.\, A(\overline{x}) \leftrightarrow S,$$

where $A(\overline{x})$ and H are as above and S is the specified relation. The proof of equivalence is conducted in an appropriate first-order theory containing axioms

$$
\begin{array}{ll}
\vdash Prog[Base] \leftrightarrow Spec[Base] & \\
\vdots & symbolic\ evaluation \\
\vdash Formula_1 \leftrightarrow Formula_2 & \\
\vdots & simplification \\
\vdash true & \\
 & tautology
\end{array}
$$

Figure 1. Schematic base case in verification.

and induction principles for recursively defined data-types. For instance, to verify the *subset* program with respect to the specification

$$\forall x.\, member(x, i) \rightarrow member(x, j),$$

we prove, in a standard theory of lists,

$$\forall i, j.\, subset(i, j) \leftrightarrow (\forall x.\, member(x, i) \rightarrow member(x, j)) \tag{6}$$

from the definitions (4) and (5). Such proofs underline our synthesis approach.

3.2. Planning Logic Program Verification Proofs

To illustrate verification proof planning, we verify *subset*; that is, we prove that the logic program (5) verifies conjecture (6). Recursion analysis suggests structural induction on i. The base case is

$$\vdash subset(nil, j) \leftrightarrow (\forall x.\, member(x, nil) \rightarrow member(x, j)).$$

Symbolic evaluation using the base cases of *subset* and *member* yields

$$\vdash true \leftrightarrow (\forall x.\, false \rightarrow member(x, j)),$$

which simplifies to *true*. The sequence of symbolic evaluation, simplification, and tautology checking is typical of the base cases of verification proofs (see Figure 1).

The annotated step case for induction on i is

$$
\begin{array}{l}
subset(t, j) \leftrightarrow (\forall x.\, member(x, t) \rightarrow member(x, j)) \vdash \\
subset\Big(\boxed{h :: \underline{t}}, j\Big) \leftrightarrow \Big(\forall x.\, member\Big(x, \boxed{h :: \underline{t}}\Big) \rightarrow member(x, j)\Big).
\end{array} \tag{7}
$$

To ripple, we need the wave rules

$$subset\Big(\boxed{H :: \underline{T}}, L\Big) :\Rightarrow \boxed{member(H, L) \wedge \underline{subset(T, L)}} \tag{8}$$

$$member\left(X, \boxed{H :: \underline{T}}\right) :\Rightarrow \boxed{X = H \vee \underline{member(X,T)}} \tag{9}$$

$$\boxed{P \vee \underline{Q}} \rightarrow R :\Rightarrow \boxed{P \rightarrow R \wedge \underline{Q \rightarrow R}} \tag{10}$$

$$\forall x. \boxed{P \wedge \underline{Q}} :\Rightarrow \boxed{(\forall x. P) \wedge \underline{\forall x. Q}} \tag{11}$$

$$\boxed{P \wedge \underline{Q}} \leftrightarrow \boxed{P \wedge \underline{R}} :\Rightarrow Q \leftrightarrow R \tag{12}$$

of which the first two are based on (4) and (5) and the remaining three on theorems of first-order logic. The latter are called *logical* wave rules (see Section 5.1).

The rippling of the induction conclusion in *subset* example consists of applying (8)–(11)

$$subset\left(\boxed{h :: \underline{t}}, j\right) \leftrightarrow \forall x. member\left(x, \boxed{h :: \underline{t}}\right) \rightarrow member(x,j)$$

$$\boxed{member(h,j) \wedge \underline{subset(t,j)}} \leftrightarrow \forall x. member\left(x, \boxed{h :: \underline{t}}\right) \rightarrow member(x,j)$$

$$\boxed{member(h,j) \wedge \underline{subset(t,j)}} \leftrightarrow \forall x. \boxed{x = h \vee \underline{member(x,t)}} \rightarrow member(x,j)$$

$$\boxed{member(h,j) \wedge \underline{subset(t,j)}} \leftrightarrow$$
$$\forall x. \boxed{x = h \rightarrow member(x,j) \wedge \underline{member(x,t) \rightarrow member(x,j)}}$$

$$\boxed{member(h,j) \wedge \underline{subset(t,j)}} \leftrightarrow$$
$$\boxed{(\forall x. x = h \rightarrow member(x,j)) \wedge \underline{\forall x. member(x,t) \rightarrow member(x,j)}}$$

and simplifying the wave front on the right-hand side

$$\boxed{member(h,j) \wedge \underline{subset(t,j)}} \leftrightarrow$$
$$\boxed{member(h,j) \wedge \underline{\forall x. member(x,t) \rightarrow member(x,j)}}$$

so that (12) can be applied:

$$subset(t,j) \leftrightarrow \forall x. member(x,t) \rightarrow member(x,j).$$

Strong fertilization completes the proof plan. The sequence of rippling both sides of the equivalence, applying a wave rule that removes the wave fronts, and strong fertilizing is typical of the step cases of verification proofs (see Figure 2).

3.3. From Verification to Synthesis

Middle-out reasoning can be used to turn verification proof planning into synthesis by planning the verification of a program while leaving the program unknown. We start with a program whose body is represented with a meta-variable. In the

$$Prog[Arg] \leftrightarrow Spec[Arg] \vdash Prog\left[\boxed{Constr[\underline{Arg}]}\right] \leftrightarrow Spec\left[\boxed{Constr[\underline{Arg}]}\right]$$

$$\vdots \qquad \textit{ripple under equivalence}$$

$$Prog[Arg] \leftrightarrow Spec[Arg] \vdash \boxed{Front[\underline{Prog[Arg]}]} \leftrightarrow \boxed{Front[\underline{Spec[Arg]}]}$$

$$\vdots \qquad \textit{final ripple}$$

$$Prog[Arg] \leftrightarrow Spec[Arg] \vdash Prog[Arg] \leftrightarrow Spec[Arg]$$

$$\textit{strong fertilization}$$

Figure 2. Schematic step case in verification.

course of planning, the variable becomes instantiated to a program. The planning thus results both in an instantiation of the program body and a plan for the verification of that program. If the plan executes successfully, the synthesized program is partially correct and complete.

Representing the program body with a meta-variable entails a loss of information, which affects the proof planning. In the verification proof above, there were a number of steps that depended on the program, but also some that did not. In particular, the symbolic evaluation of *subset* in the base case and the ripple with the *subset* wave rule (8) depend on the program. The main difference between verification and synthesis planning is that in verification, the two types of steps tend to be interleaved, whereas in synthesis, the part of the proof that does not depend on the program is planned first, and any step that does is postponed as long as possible. This is because, in synthesis, any step that depends on the program partially instantiates, i.e., commits the program. Postponing such steps is a least commitment strategy.

To illustrate this, we redo the step case of the verification proof in Section 3.2, omitting any steps that depend on the program. This rules out rippling with the *subset* wave rule (8). The rippling progresses as follows, using wave rules (9)–(11) and unblocking.

$$subset\left(\boxed{h :: \underline{t}}, j\right) \leftrightarrow \forall x. \boxed{x = h \vee \underline{member(x,t)}} \rightarrow member(x,j)$$

$$subset\left(\boxed{h :: \underline{t}}, j\right) \leftrightarrow$$

$$\forall x. \boxed{x = h \rightarrow member(x,j) \wedge \underline{member(x,t) \rightarrow member(x,j)}}$$

$$subset\left(\boxed{h :: \underline{t}}, j\right) \leftrightarrow$$

$$\boxed{(\forall x. x = h \rightarrow member(x,j)) \wedge \underline{\forall x. member(x,t) \rightarrow member(x,j)}}$$

$$Prog[Arg] \leftrightarrow Spec[Arg] \vdash Prog\left[\boxed{Constr[\underline{Arg}]}\right] \leftrightarrow Spec\left[\boxed{Constr[\underline{Arg}]}\right]$$

$$\vdots \qquad \textit{ripple under equivalence}$$

$$Prog[Arg] \leftrightarrow Spec[Arg] \vdash Prog\left[\boxed{Constr[\underline{Arg}]}\right] \leftrightarrow \boxed{Front[\underline{Spec[Arg]}]}$$

$$\vdots \qquad \textit{weak fertilization}$$

$$Prog[Arg] \leftrightarrow Spec[Arg] \vdash Prog\left[\boxed{Constr[\underline{Arg}]}\right] \leftrightarrow \boxed{Front[\underline{Prog[Arg]}]}$$

$$\textit{appeal to program}$$

Figure 3. Schematic step case in synthesis.

$$\vdash Prog[Base] \leftrightarrow Spec[Base]$$

$$\vdots \qquad \textit{symbolic evaluation, simplification}$$

$$\vdash Prog[Base] \leftrightarrow Formula$$

$$\textit{appeal to program}$$

Figure 4. Schematic base case in synthesis.

$$subset\left(\boxed{h :: \underline{t}}, j\right) \leftrightarrow$$
$$\boxed{member(h, j) \land \underline{\forall x.\ member(x, t) \rightarrow member(x, j)}}$$

The lack of wave rule for *subset* now prevents us from further rippling. However, we can apply weak fertilization: We use the induction hypothesis (7) as a rewrite rule. This yields

$$subset\left(\boxed{h :: \underline{t}}, j\right) \leftrightarrow \boxed{member(h, j) \land \underline{subset(t, j)}}.$$

Now we have applied all possible steps that do not depend on the program. In fact, the residual conjecture is precisely the part of the proof that in a verification, would have been proved using the *subset* wave rule (8). It is thus the step clause of our program. By appealing to the as-yet uninstantiated program, we commit it to correspond to this residual conclusion. The details of this process are presented below. Figure 3 shows schematically how a typical step case of a synthesis proof progresses: The specification side of the induction conclusion is rippled until weak fertilization is possible, and the proof is completed by appealing to the program.

The base case is similar. In synthesis, only the specification side is symbolically evaluated and simplified, and the residual conclusion corresponds to part of the program (see Figure 4).

3.4. An Example Synthesis

We synthesize the program we verified in Section 3.2. The *subset* program is now

$$\forall i, j.\, subset(i, j) \leftrightarrow \mathcal{P}(i, j),$$

where $\mathcal{P}$ is the meta-variable that represents the program body. We again do structural induction on i. The type of induction immediately determines the recursive structure of the program. Each induction scheme is associated with the corresponding recursive structure, and the program body is unified (see Section 4.2) with this structure

$$\begin{aligned} \forall i, j.\, subset(i, j) \;\leftrightarrow\; & i = nil \wedge \mathcal{B}(j) \vee \\ & \exists h, t.\, i = h :: t \wedge \mathcal{S}(h, t, j). \end{aligned} \tag{13}$$

The base case for induction on i is

$$\vdash subset(nil, j) \leftrightarrow (\forall x.\, member(x, nil) \rightarrow member(x, j)).$$

Symbolic evaluation using the base case of *member* and simplification yield

$$\vdash subset(nil, j) \leftrightarrow true. \tag{14}$$

We are now left with what will become the base case of the program. By appealing to the as-yet uninstantiated program definition, we complete the base case of the proof and at the same time instantiate the base case of the program. This is done by the `synthesis` method. To appeal to the program (13), the `synthesis` method instantiates it appropriately and simplifies it

$$\begin{aligned} subset(nil, j) \;\leftrightarrow\; & nil = nil \wedge \mathcal{B}(j) \vee \\ & \exists h', t'.\, nil = h' :: t' \wedge \mathcal{S}(h', t', j) \\ subset(nil, j) \;\leftrightarrow\; & \mathcal{B}(j). \end{aligned} \tag{15}$$

The conclusion (14) and the program (15) are unified (see Section 4.2), which yields the instantiation $\lambda u.\, true$ for $\mathcal{B}$ and completes the base case. The (normalized) partially instantiated program so far is thus

$$\begin{aligned} \forall i, j.\, subset(i, j) \;\leftrightarrow\; & i = nil \wedge true \vee \\ & \exists h, t.\, i = h :: t \wedge \mathcal{S}(h, t, j). \end{aligned} \tag{16}$$

In Section 3.3, we showed that the residual conclusion in the step case is

$$subset\left(\boxed{h :: \underline{t}}, j\right) \leftrightarrow \boxed{member(h, j) \wedge \underline{subset(t, j)}}. \tag{17}$$

We must establish that this follows from the program definition. The `synthesis` method instantiates (16) appropriately and simplifies it

$$\begin{aligned} subset(h :: t, j) &\leftrightarrow h :: t = nil \wedge true \vee \\ &\qquad \exists h', t'. h :: t = h' :: t' \wedge \mathcal{S}(h', t', j) \\ subset(h :: t, j) &\leftrightarrow \mathcal{S}(h, t, j). \end{aligned} \tag{18}$$

Unifying (see Section 4.2) the conclusion (17) (with annotations removed) and the program (18) instantiates $\mathcal{S}$ with $\lambda u, v, w.\ member(u, w) \wedge subset(v, w)$ and completes the step case. We get the (normalized) fully instantiated program

$$\begin{aligned} \forall i, j.\ subset(i, j) &\leftrightarrow i = nil \wedge true \vee \\ &\qquad \exists h, t. i = h :: t \wedge member(h, j) \wedge subset(t, j). \end{aligned}$$

3.5. Auxiliary Syntheses

In the course of a synthesis, we need to prevent a meta-variable from becoming instantiated with a program body that violates the definition of pure logic programs (see Section 3.1). Instead of directly checking the instantiation, *Periwinkle* parses the program on completion of a synthesis, marking any subformulae that violate the syntactic restrictions on pure logic programs. The universal closure of each such subformula is taken as the specification for an auxiliary synthesis. In the initial program, any subformula for which an auxiliary synthesis was run is substituted with a call to the corresponding auxiliary predicate, and all auxiliary predicates are added to the program. Note that an auxiliary synthesis may require further auxiliary syntheses. Though the process is not guaranteed to terminate, non-termination has not been a problem in practice. An example where an auxiliary synthesis is necessary is the specification

$$\forall m, l.\ max(m, l) \leftrightarrow member(m, l) \wedge (\forall x.\ member(x, l) \rightarrow x \leqslant m),$$

which states that m is the maximum element of l. The synthesized program is

$$\begin{aligned} \forall m, l.\ max(m, l) &\leftrightarrow l = nil \wedge false \vee \\ &\qquad \exists h, t. l = h :: t \wedge (m = h \wedge (\forall x.\ member(x, t) \rightarrow \\ &\qquad\qquad x \leqslant m) \vee h \leqslant m \wedge max(m, t)). \end{aligned}$$

The subformula $\forall x.\ member(x, t) \rightarrow x \leqslant m$ violates the definition of pure logic programs, since it contains a universal quantifier and an implication. The auxiliary specification is

$$\forall m, l.\ aux(m, l) \leftrightarrow (\forall x.\ member(x, l) \rightarrow x \leqslant m),$$

which states that m is greater than any element of l. The final program is

$$\begin{aligned}\forall m, l.\, max(m,l) \;\leftrightarrow\;& l = nil \wedge false \vee \\ & \exists h, t.\, l = h :: t \wedge (m = h \wedge aux(m,t) \vee \\ & \qquad h \leqslant m \wedge max(m,t)) \\ \forall m, l.\, aux(m,l) \;\leftrightarrow\;& l = nil \wedge true \vee \\ & \exists h, t.\, l = h :: t \wedge h \leqslant m \wedge aux(m,t).\end{aligned}$$

3.6. Related Work in Program Synthesis

Most program synthesis approaches have originated in the field of functional programming. There has, however, been increased interest in adapting these to logic program synthesis. For a detailed overview of logic program synthesis, see [20].

Fribourg [21] and Wiggins [42] both adapt the *proofs-as-programs* approach to logic program synthesis. Fribourg, however, uses $\forall\exists$ specifications. The synthesized programs are thus not truly relational. Wiggins develops a synthesis logic for relational program synthesis with a decidability operator, implemented in a system called *Whelk*. Both systems are interactive. LOPS [4, 5] transforms first-order specifications into logic clauses. LOPS also uses $\forall\exists$ specifications and thus is not really relational. The semi-automatic system of Lau and Prestwich [32, 33] is based on unfold/fold transformations of logic programs. Our approach synthesizes truly relational programs, like Wiggins and Lau and Prestwich, and unlike Fribourg and LOPS. It is also fully automatic. A more detailed comparison of our approach with those of Wiggins and Lau and Prestwich follows.

The emphasis in *Whelk* [42] is to develop a logic in which relational programs can be synthesized via proofs-as-programs-style extraction. Thus, in the *Whelk* system, synthesis takes place at the object level, not the meta-level, and correctness and executability are ensured in the object-level logic. By contrast, we synthesize and ensure executability at the meta-level, while establishing partial correctness and completeness by a verification proof at the object level. The difference between the two approaches lies in emphasis. While the *Whelk* project focuses more on the logical issues of logic program synthesis, we have put more emphasis on automation. We have therefore chosen as our object-level logic a well-understood formal system, i.e., many-sorted first-order logic with induction, and have taken a perhaps pragmatic approach by using middle-out reasoning for synthesis and by ensuring executability using extralogical means. In the *Whelk* project, on the other hand, a special logic with a decidability operator was developed to synthesize guaranteed executable programs, while automation was a secondary priority.

The proof planning system *CLAM* is currently being adapted to plan proofs in the *Whelk* logic. The techniques developed here will be directly applicable, in particular middle-out induction (see Section 4) and extensions to rippling (see Section 5). On the other hand, results in the *Whelk* logic could be used to ensure executability at the meta-level without extra-logical means, thus improving our handling of auxiliary syntheses.

The system of Lau and Prestwich [32, 33] is semi-automatic and unfold/fold-based. It synthesizes partially correct, but not necessarily complete programs. It solves a synthesis problem by bringing it into a normal form and decomposing it top-down into subproblems until the subproblems are easily solved. The program is then composed bottom-up from the solutions of the subproblems. User interaction is required to limit the search space by specifying the desired recursive calls of the program and by deciding which subproblems to solve. The main strategies are definition, implication and matching. The *definition* strategy selects a definition and uses the *if* part of the definition to unfold and the *only if* part to fold. The *implication* strategy exploits known recursive implications. The *match* strategy solves trivial folding problems.

The *subset* example is taken from [32]. It is thus a good candidate for comparison. The inputs to the system of Lau and Prestwich are the *subset* specification, the definition of *member*, and a goal specifying the initial unfold/fold problem. The definition strategy is applied twice, which results in two subproblems. The first is solved with the implication, the second with the match strategy. There are no remaining subproblems, and the solution can be composed. This involves the actual unfolding and folding, interleaved with steps to bring intermediate formulae into various types of normal forms. Setting the initial fold problem corresponds to selecting the type of induction. The initial unfolding then corresponds to induction, and the last folding to fertilization. The remaining decomposition, normalization, and composition steps correspond to rippling. In fact, we believe that approach of Lau and Prestwich could be improved by exploiting rippling to guide the folding. This would obviate the need for normal forms.

Lau and Prestwich synthesize partially correct, but not necessarily complete programs, whereas we insist on both partial correctness and completeness. Not requiring completeness has the advantage that the body of the program being synthesized can be strengthened. Although strengthening allows greater flexibility in synthesis, it also increases the search space, which in Lau and Prestwich's work translates into a need for user interaction. Nevertheless, our approach could well benefit from the strategies of Lau and Prestwich that strengthen formulae to allow folding. This may well be essential when synthesizing larger, more complex programs, or synthesizing programs from partial specifications.

4. Middle-Out Induction

Determining the appropriate type of induction for a given conjecture is a difficult task. The most widely used technique is *recursion analysis* (see Section 2). However, recursion analysis works poorly in the presence of existential quantifiers, which arise in $\forall\exists$ specifications of functions. This is because the appropriate induction scheme is bound to the recursion scheme of the witnessing function – which is precisely what we want to synthesize and therefore do not know. Using an inappropriate induction scheme may make it difficult to find a proof and may lead to an unintuitive or inefficient program.

An example where recursion analysis breaks down is *quotient-remainder*

$$\forall x, y.\, \exists q, r.\, x \neq 0 \rightarrow q \times x + r = y \wedge r < x.$$

Only x and y are available as induction variables, and given the standard definitions of $\times$, $+$, and $<$, recursion analysis cannot find the appropriate induction, which is induction on y, where the induction term is $y + x$.

Recursion analysis works better for the relational conjectures in our approach. The conjecture for a quotient-remainder relation qr is

$$\forall x, y, q, r.\, qr(x, y, q, r) \leftrightarrow q \times x + r = y \wedge r < x,$$

where qr is undefined. Since the conjecture is universally closed, we can choose any of x, y, q, and r as induction variables, not only x and y. Hence, recursion analysis stands a better chance of success. For this conjecture, recursion analysis does suggests an appropriate induction: one-step structural induction on q.

However, recursion analysis is always limited to finding a type of induction based on the recursion schemes present in the specification or in given lemmas. Even for relational conjectures, the recursion of the program may not be among them. An example of this is the conjecture

$$\forall x.\, even(x) \leftrightarrow (\exists y.\, y \times s(s(0)) = x),$$

where *even* is undefined. The natural recursion of the program is two-step recursion, which is not suggested by the standard definition of $\times$. Therefore, we need a more powerful technique.

Middle-out reasoning can provide such a technique. It can be used to postpone the first, crucial step in the planning of inductive proofs, namely the selection of an induction. This was first suggested by Bundy et al. [12], but had not been elaborated or implemented. Using meta-variables, we can set up a schematic step case representing many possible inductions. This is achieved by using meta-variables to represent constructors applied to potential induction variables in the induction conclusion. We can then ripple this schematic step case. The application of wave rules successively instantiates the meta-variables. Once fertilization has taken place, the meta-variables are fully instantiated and correspond to a type of induction. We do need to ensure that the induction is a valid one, i.e., that the induction ordering is well founded. The most general approach would be to prove that the order is in fact well founded. This is a difficult task, undecidable in general. We currently use a simpler approach, which is to check whether the ordering is among a set of orderings known to be well founded. Once we have determined that the induction is valid, we can set up the corresponding base cases and complete the proof using standard proof planning methods.

Middle-out induction has two main advantages over recursion analysis: first, it is a more general approach. It can find an appropriate induction even in cases where recursion analysis fails. Second, recursion analysis essentially performs a look-ahead into the rippling process, whereas middle-out induction requires no such look-ahead. However, there are two problems to overcome in middle-out induction: It requires some kind of higher-order unification, and rippling is no longer terminating. These problems are discussed below.

4.1. An Example Synthesis with Middle-Out Induction

We present a variation of the *even* specification above.[2] The conjecture is

$$\forall x.\, even(x) \leftrightarrow (\exists y.\, double(y) = x),$$

where *double* is defined as

$$double(0) = 0$$
$$\forall x.\, double(s(x)) = s(s(double(x))).$$

The wave rules for *double* and the replacement axiom for s are

$$double\left(\boxed{s(\underline{U})}\right) :\Rightarrow \boxed{s(s(\underline{double(U)}))} \tag{19}$$

$$\boxed{s(\underline{U})} = \boxed{s(\underline{V})} :\Rightarrow U = V. \tag{20}$$

The schematic step case is

$$even(x) \leftrightarrow (\exists y.\, double(y) = x) \vdash$$
$$even\left(\boxed{C(\underline{x})}\right) \leftrightarrow \left(\exists y.\, double(y) = \boxed{C(\underline{x})}\right), \tag{21}$$

where C is the meta-variable standing for the constructor applied to the potential induction variable, and the dashed boxes indicates potential wave fronts; that is, C may be instantiated to some function that becomes the wave front or to the identity function $\lambda x.\, x$. The latter means that there is no wave front. Initially, no wave rule applies. To make a wave rule applicable, we need to introduce a case split on the existential variable y. This is done by the `unrolling` method, which is presented in detail in Section 5.2. Unrolling on y yields

$$even\left(\boxed{C(\underline{x})}\right) \leftrightarrow$$
$$\boxed{double(0) = C(x) \vee \underline{\exists y'.\, double\left(\boxed{s(\underline{y'})}\right) = \boxed{C(\underline{x})}}}.$$

Applying wave rule (19) results in

$$even\left(\lceil \mathcal{C}(\underline{x}) \rceil\right) \leftrightarrow$$

$$\boxed{double(0) = \mathcal{C}(x) \vee \exists y'. \underline{\boxed{s(s(\underline{double(y')}))} = \lceil \mathcal{C}(\underline{x}) \rceil}}.$$

Applying wave rule (20) twice then results in

$$even\left(\boxed{s\left(s\left(\lceil \underline{C''(\underline{x})} \rceil\right)\right)}\right) \leftrightarrow$$

$$\boxed{double(0) = s\Big(s\big(C''(x)\big)\Big) \vee \exists y'. \underline{double(y') = \lceil C''(\underline{x}) \rceil}}.$$

We can simplify and weak fertilize, i.e., apply the induction hypothesis (21) as a rewrite rule. This step yields

$$even(s(s(x))) \leftrightarrow even(x).$$

Weak fertilization instantiates C to $\lambda u.\, s(s(u))$. The step case and the base cases are now completed, as described in Section 3.

4.2. UNIFICATION

Since we use higher-order meta-variables in our middle-out reasoning, we are confronted with the problem of higher-order unification, which is only semi-decidable. Moreover, there is no unique most-general unifier of higher-order terms. When using higher-order terms, therefore, one either accepts this fact and uses, for instance, the procedure of Huet [25], combined with backtracking over or selection of possible unifiers, or one uses a restricted subset of higher-order terms with tractable unification, e.g., *higher-order patterns*. The former approach has been taken, for instance, by Hesketh [24] and Ireland [27]. The latter approach is taken here.

Higher-order patterns [37, 38] are expressions whose free variables have no arguments other than bound variables. Formally, following [38],

> "a term t in β-normal form is called a (*higher-order*) *pattern* if every free occurrence of a variable F is in a subterm $F(u_1, \ldots, u_n)$ of t such that each u_i is η-equivalent to a bound variable and the bound variables are distinct."

Higher-order patterns are akin to first-order terms in that unification is decidable and there exists a unique most-general unifier of unifiable terms.

We have restricted ourselves to higher-order patterns for the terms in which we use meta-variables because they fall naturally into the class of higher-order patterns. For synthesis proper, we are creating programs that represent relations and that are therefore developed in the context of a collection of universally bound variables. The distinctness requirement is already satisfied by the definition of pure logic programs. Thus, what we start out with as our program is already a higher-order pattern. Any step that further instantiates the higher-order pattern does so via unification with another higher-order pattern. For middle-out induction, we use meta-variables to represent the constructor function applied to the induction variable. Since the variable on which we induce must be universally bound to begin with, the expressions we obtain are again higher-order patterns. Furthermore, the instantiation of the meta-variables occurs via the application of wave rules, which are also higher-order patterns.

4.3. A More General Representation of the Step Case

The representation of the schematic step case used above does not cover more complex induction schemes where the induction term for a variable refers also to other variables. This is the case, for instance, in the quotient-remainder example

$$\forall x, y, q, r.\, qr(x, y, q, r) \leftrightarrow q \times x + r = y \wedge r < x,$$

where the induction term for y is $y + x$. Above, we represented the induction term for a variable with a meta-variable applied to the variable, e.g., $\mathcal{D}(y)$ for y. Since the potential induction variables are bound, the instantiation of $\mathcal{D}(y)$ cannot refer to the variables x, q, or r. To allow this, we must generalize the representation of the schematic step case by representing the induction term for a potential induction variable as an application of a meta-variable to *all* potential induction variables. Thus, the induction term for y is represented as $\mathcal{D}(x, y, q, r)$, which, properly annotated, becomes $\boxed{\mathcal{D}(x, \underline{y}, q, r)}$. The schematic step case is then

$$
\begin{aligned}
& qr(x, y, q, r) \leftrightarrow q \times x + r = y \wedge r < x \vdash \\
& qr\left(\boxed{\mathcal{C}(\underline{x}, y, q, r)}, \boxed{\mathcal{D}(x, \underline{y}, q, r)}, \boxed{\mathcal{E}(x, y, \underline{q}, r)}, \boxed{\mathcal{F}(x, y, q, \underline{r})}\right) \leftrightarrow \\
& \quad \boxed{\mathcal{E}(x, y, \underline{q}, r)} \times \boxed{\mathcal{C}(\underline{x}, y, q, r)} + \boxed{\mathcal{F}(x, y, q, \underline{r})} = \boxed{\mathcal{D}(x, \underline{y}, q, r)} \wedge \\
& \quad \boxed{\mathcal{F}(x, y, q, \underline{r})} < \boxed{\mathcal{C}(\underline{x}, y, q, r)}.
\end{aligned}
$$

While dealing with this representation is not a problem for *Periwinkle*, it is not particularly fit for human consumption. The implementation supports both the simpler representation in Section 4.1 and the more complex representation here.

4.4. Controlling Rippling

Two of the main advantages of rippling are that it gives a tight control on rewriting and that it terminates. The termination proof [14, 3] makes some restrictions, i.e., existential rippling (see Section 5.2) and meta-variables are excluded, precisely because they can lead to non-termination. Since middle-out synthesis and induction require meta-variables, we must contend with the possibility of non-termination and devise strategies to avoid it.

Non-termination is in fact more likely than not in rippling in middle-out induction. In terms of the rippling search tree in the schematic step case, where each node corresponds to the application of a wave rule, we can differentiate between two basic types of non-termination: non-termination in success branches and non-termination in failure branches.

Non-termination in success branches can be avoided by distinguishing between speculative and non-speculative steps. Applying a wave rule to potential wave fronts only, for instance, is a speculative step. Fertilization and applying a wave rule to at least one definite wave front are non-speculative steps. By preferring non-speculative to speculative steps, non-termination on success branches can be avoided. However, this does not avoid non-termination in failure branches. If there were always at least one success branch in the rippling search tree, breadth-first search would solve the problem. Unfortunately, however, this is not the case. A simple example of a rippling search tree with failure branches only is a variant of the associativity of plus[3]

$$\forall x.\, x + (x + x) = (x + x) + x.$$

To avoid non-termination in failure branches, we allow only one speculative step, which can be a speculative ripple or an unrolling step (see Section 5.2), and then ripple while trying to fertilize as soon as possible. This does mean that *Periwinkle* cannot find a proof for theorems that depend on more than one speculative ripple. This situation is in fact rare, so that it does not appear to be a severe limitation.

4.5. Related Work in Selection of Induction Schemes

There has been little work on techniques to select induction schemes beyond recursion analysis, except within the framework of the Inka theorem-prover [8], a theorem prover based on resolution and rippling with destructor-style induction.

The synthesis system of Biundo [6] applies a heuristic called *most-nested function* to specifications in the form of Skolemized $\forall\exists$ formulae. A most-nested function is one that occurs at an innermost position in the specification. The recursive arguments of the most-nested function are selected as the induction variables, and the recursion of the most-nested function as the type of induction. The recursive arguments of the most-nested function should also be among the set of variables that are arguments of the Skolem function. The type of recursion of the program will then correspond to the type of recursion of the most-nested function. While this heuristic is sufficient for some simple examples, it performs as poorly as recursion analysis in more complex examples. For examples, for the (Skolemized) quotient-remainder specification from [7]

$$\begin{aligned}\forall x, y.\, y \neq 0 \rightarrow\ & plus(times(car(f(x,y)), x), cdr(f(x,y))) = y\ \wedge \\ & cdr(f(x,y)) < x,\end{aligned}$$

the heuristic selects structural induction on y. While the system does find a proof, the resulting program is unintuitive and inefficient.

Protzen [39] presents a calculus for constructing induction proofs without committing to an induction scheme at the outset. The ideas are closely related to our middle-out reasoning approach, first presented in [31], except that they are couched within a destructor style setting and a specialized proof system. The idea is to generate induction hypotheses by rewriting the induction conclusion until all structure introduced by the rewriting is in *tolerable positions*, i.e., surrounds the conclusion or potential induction variables. Tolerable positions basically correspond to our meta-variables. Additionally, Protzen's calculus forces the induction hypotheses to be smaller relative to a well-founded ordering. The approach has not, to the best of our knowledge, been fully implemented yet. We have two concerns with Protzen's techniques. First, in our experience, the search space in such proofs is very large and rather sensitive to how the rewriting is constrained. Hence, carefully designed search control is essential, and it is likely that, in practice, Protzen will have to introduce heuristics to restrict the search to have a usable system. Second, it is not clear whether the approach applies to universally quantified conjectures only or also to existentially quantified and synthesis conjectures.

Hutter [26] suggests a technique to select induction schemes for $\forall\exists$ formulae that exploits the close relationship between the induction variables, the instantiation of existential variables, and the type of induction. Instead of selecting induction variables and the type of induction and then finding instantiations of existential variables, Hutter picks induction variables and instantiations of existential variables, leaving the type of induction to be determined in the course of the proof. Hutter's approach involves two steps: first, the selection of an induction variable and an existential variable and second, the selection of the induction scheme. The selection of the pair of variables is done in a preprocessing step bearing some similarity to recursion analysis. First, all available wave rules are abstracted in that the only information retained is the dominating function symbol and the direction in which the wave front moves – up, down, or across. These abstracted rules are called *labeled fragments*. The term tree of the conjecture is then searched to find a path of labeled fragments such that all instances of a universal and an existential variable are connected, but none of the fragments overlap. Such a path ensures that there is wave rule that can move a wave front in the desired direction at every relevant node. It does not consider the actual form of the wave front, however. Once the variables have been selected, the actual proof is carried out. In the base case, the existential variable is instantiated to the base of the corresponding type. Then, symbolic evaluation is applied. The remaining formula is assumed as the condition of the base case, which completes its proof. The negation of this formula becomes the condition of the step case. In the step case, the existential variable is again instantiated, now to the compound case of the type, and the conclusion is rippled. Once the rippling has terminated, the structure that has accumulated around the induction variable determines its predecessor. As in our approach, the well-foundedness of the induction order remains to be established. To the best of our knowledge, the approach has not yet been implemented.

Hutter's approach and ours are related. Both rely on the rippling of the step case to determine the type of induction. Both require a certain amount of search, Hutter's in the preprocessing step, ours in the rippling. The main difference lies in the fact that Hutter's approach is divided into two steps. The preprocessing corresponds to a lookahead into the rippling, albeit a simplified version. The trade-off between our one-step and Hutter's two-step approach is thus that Hutter's does some of the search in a simplified setting, which reduces the amount of search in the actual rippling, but involves some duplication of effort. We search in the actual rippling, which is more expensive, but we have no duplication of effort. Finally, the preprocessing step of Hutter simply fails if a lemma is missing, since it cannot find a path. This failure would pose a serious problem in proofs requiring propositional wave rules, when, as in our system, these wave rules are generated on demand.

5. Extensions to Rippling

Synthesizing logic programs is a new application of proof planning and poses new problems to the proof planner. While many of them are specific to program synthesis, some of them can occur in proof planning in general. Methods developed to solve these more general problems are presented together in this section.

5.1. Generating Logical Wave Rules

Initially, *Periwinkle* used a library of around sixty wave rules based on schematic lemmas about logical connectives. Rules (10) and (11) in Section 3.4 are examples of such rules. Considering the large number of such wave rules, it would be preferable for *Periwinkle* to recognize the need for one and generate it on demand. A method that does this has been implemented for a large subclass of logical wave rules, i.e., wave rules expressed in terms of propositional connectives only. Wave rule (10) is propositional, whereas wave rule (11) is not, since it involves quantifiers.

The idea underlying the generation of propositional wave rules is that we can conjecture a partially specified lemma that gives rise to the desired wave rule. We then try to fill in the missing part of the lemma by generating the truth table for that part and finding a formula that satisfies that truth table. In the conclusion

$$\ldots \leftrightarrow \neg \boxed{x = h \vee \underline{\mathit{member}(x,t)}},$$

for instance, the rippling is blocked. To ripple the right-hand side further, we need a wave rule that pushes the negation down over the disjunction. The wave rule we want is thus of the form

$$\neg\boxed{P \vee \underline{Q}} :\Rightarrow \boxed{\mathcal{F}(\underline{\neg Q})},$$

based on a lemma

$$\neg(P \vee Q) \leftrightarrow \mathcal{F}(\neg Q),$$

where $\mathcal{F}$ represents the unknown part. We need to find an expression $\mathcal{F}$ with the same truth values as $\neg(P \vee Q)$ containing a subterm $\neg Q$. We proceed as follows: first, we try the simple cases: $\neg Q$ itself and its negation $\neg\neg Q$. This approach fails, and we create a set of candidate expressions whose top-level connective is some binary connective, with $\neg Q$ as its first argument and an unknown expression E as its second argument. In this example, we consider only conjunction and disjunction, though the implementation also considers equivalence and implication. We construct the candidate expressions and derive the truth tables for the second argument of the top-level connective. The truth table for E in $\neg Q \vee E$ is

P	Q	$\neg(P \vee Q)$	$\neg Q$	E
T	T	F	F	F
T	F	F	T	X
F	T	F	F	F
F	F	T	T	T/F

and the truth table for E in $\neg Q \wedge E$ is

P	Q	$\neg(P \vee Q)$	$\neg Q$	E
T	T	F	F	T/F
T	F	F	T	F
F	T	F	F	T/F
F	F	T	T	T

In the column for E, T indicates that E must be true for the given values of the variables P and Q, F that E must be false, and T/F that E may be either true or false. The symbol X indicates a conflict; no possible value is logically consistent. Any formula with a conflict is discarded. We try to complete the surviving candidates by finding a variable or a negation of a variable that satisfies the derived truth table. In this example, the truth table for $\neg Q \wedge E$ is satisfied by $\neg P$. If no (negated) variable had satisfied the truth table for E, we would have constructed more complex expressions using binary connectives. To cut down on the search space, we restrict the first argument of any binary connective to a propositional variable or its negation, but allow the second argument to be further expanded.

This straightforward approach to generating propositional wave rules sufficed to generate all propositional wave rules in our original database. However, it is not clear whether it is complete. The problem lies in the syntactic restrictions that were made to cut down the search space, and the question is whether there are wave rules that cannot be expressed in the restricted syntactic form.

5.2. Existential Rippling and Unrolling

Rippling as portrayed in Section 2 applies to theorems containing universal quantifiers only, and cannot cope with existential quantifiers. To allow a wave rule to apply to existential variables, we introduce existential versions of wave rules. In essence, what an existential wave rule does is to partially instantiate an existential variable to a wave term. A simple wave rule of the form

$$F\left[\boxed{S[\underline{U}]}\right] :\Rightarrow \boxed{T[\underline{F[U]}]}$$

can be turned into an existential wave rule

$$\exists \boxed{u}.\, G\left[F\left[\boxed{u}\right]\right] :\Rightarrow \exists \boxed{u'}.\, G\left[\boxed{T\left[F\left[\boxed{u'}\right]\right]}\right].$$

To indicate that existential variables can be instantiated to wave terms, the variables are annotated with potential wave fronts, represented as dashed boxes.

There is a problem when applying existential rippling in logic program synthesis. Wave rules used to rewrite subexpressions of equivalences must be based on equivalence or equality. Unfortunately, existential versions of such wave rules are not necessarily equivalence-preserving. An example where the existential wave rule is not equivalence-preserving is the synthesis conjecture that a list k occurs at the back of a list l

$$\forall k, l.\, back(k, l) \leftrightarrow \exists x.\, app(x, k, l).$$

Applying structural induction on l yields the step case

$$\forall k.\, back\left(k, \boxed{h :: \underline{t}}\right) \leftrightarrow \exists \boxed{x}.\, app\left(\boxed{x}, k, \boxed{h :: \underline{t}}\right). \tag{22}$$

Now, we would like to apply the wave rule

$$app\left(\boxed{H_1 :: \underline{T_1}}, L, \boxed{H_2 :: \underline{T_2}}\right) :\Rightarrow \boxed{H_1 = H_2 \wedge \underline{app(T_1, L, T_2)}} \tag{23}$$

in its existential version

$$\exists \boxed{x}.\ app\left(\boxed{x}, L, \boxed{H :: \underline{T}}\right) :\Rightarrow$$
$$\boxed{\exists \lceil x_1 \rceil . \exists \lceil x_2 \rceil . \lceil x_1 \rceil = H \wedge \underline{app\left(\lceil x_2 \rceil, L, T\right)}}. \qquad (24)$$

While the lemma underlying (23)

$$\forall h_1, t_1, l, h_2, t_2.\ app(h_1 :: t_1, l, h_2 :: t_2) \leftrightarrow h_1 = h_2 \wedge app(t_1, l, t_2)$$

holds, the equivalence that would justify (24)

$$\forall l, h, t.\ (\exists x.\ app(x, l, h :: t)) \leftrightarrow (\exists x_1, x_2.\ x_1 = h \wedge app(x_2, l, t))$$

is a non-theorem (the left side is true, but the right side false for $l = h :: t$).

Thus, before we apply an existential version of wave rule, we must establish that it is equivalence-preserving. (Dis-)Proving the underlying lemma, however, is a difficult and expensive task and requires the full power of proof planning. Instead of implementing a method for equivalence-preserving existential rippling, we have developed a less expensive and more generally applicable alternative, *unrolling*.

In the step case (22), with (23) as the only available wave rule, rippling is blocked from the outset by the existentially quantified variable x. Applying wave rule (23) is clearly what is called for. For that, x must be brought into the form $\boxed{x_1 :: \underline{x_2}}$. This can be done by introducing an appropriate case split, i.e., one where one of the cases is $x = x_1 :: x_2$. The case split that lends itself is the one where x is either the empty or a composite list.

When we apply the case split, we must preserve the skeleton of the conclusion and put any additional structure introduced by the case split into a wave front. This is achieved by annotating the case split accordingly. Schematically, a case split on x in $\exists x : nat.\ P(x)$ is annotated as

$$\boxed{P[0] \vee \underline{\exists x : nat.\ P\left[\boxed{s(\underline{x})}\right]}}.$$

In the *back* example, introducing a case split on x leads to the conclusion

$$\forall k.\ back\left(k, \boxed{h :: \underline{t}}\right) \leftrightarrow$$
$$\boxed{app(nil, k, h :: t) \vee \underline{\exists x_1 . \exists x_2 .\ app\left(\boxed{x_1 :: \underline{x_2}}, k, \boxed{h :: \underline{t}}\right)}},$$

to which wave rule (23) applies, yielding

$$\forall k.\ back\left(k, \boxed{h :: \underline{t}}\right) \leftrightarrow$$
$$\boxed{app(nil, k, h :: t) \vee \underline{\exists x_1 . \exists x_2 . \boxed{x_1 = h \wedge \underline{app(x_2, k, t)}}}}.$$

The step case can be completed with a further ripple and weak fertilization.

Currently, only case splits on the structure of recursive data types are considered, both for existential and universal quantifiers. The method can be extended to cover more complex case splits.

Unrolling is a speculative step and can cause non-termination. It is therefore controlled in the same way as speculative rippling (see Section 4.4). It is worth noting the relationship between equivalence-preserving existential rippling and unrolling. Applying an equivalence-preserving existential wave rule to a conclusion

$$\exists y : nat.\, P\left(\boxed{s(\underline{x})}, y\right)$$

yields a conclusion

$$\exists y' : nat.\, \boxed{F(\underline{P(x, y')})}. \tag{25}$$

Unrolling initially yields

$$\boxed{P(x,0) \vee \underline{\exists y' : nat.P\left(\boxed{s(\underline{x})}, \boxed{s(\underline{y'})}\right)}}.$$

Because the existential rewrite was equivalence-preserving, the case $P(x, 0)$ must be false. Thus, after we simplify the wave front and apply the original version of the wave rule, we also obtain (25).

5.3. Very Weak Fertilization

The techniques presented so far are mainly devised to allow rippling to continue. In some cases, however, the blockage does not prevent further rippling, but fertilization. *Very weak fertilization* is a method that recognizes one particular type of blockage that occurs frequently when synthesizing relations from functions. It exploits the induction hypothesis in a way that takes the blockage into account. Very weak fertilization applies in particular in cases where we would like to weak fertilize, but we cannot because the corresponding side of the conclusion is an equality that is not yet fully rippled. A simple example where this occurs is the synthesis of a reverse relation from a reverse function

$$\forall k, l.\, rrev(k, l) \leftrightarrow frev(k) = l,$$

where *frev* is defined as

$$\begin{aligned} frev(nil) &= nil \\ \forall h, t.\, frev(h :: t) &= app(frev(t), h :: nil). \end{aligned}$$

From the *frev* definition, we get the wave rule

$$frev\left(\boxed{H :: \underline{T}}\right) :\Rightarrow \boxed{app(\underline{frev(T)}, H :: nil)}. \tag{26}$$

By structural induction on k, we obtain the step case

$$\forall l.\, rrev(t,l) \leftrightarrow frev(t) = l \vdash \forall l.\, rrev\left(\boxed{h :: \underline{t}}, l\right) \leftrightarrow frev\left(\boxed{h :: \underline{t}}\right) = l.$$

Applying wave rule (26) yields

$$rrev\left(\boxed{h :: \underline{t}}, l\right) \leftrightarrow \boxed{app(\underline{frev(t)}, h :: nil)} = l.$$

The rippling is now blocked. However, the wave hole on the right-hand side of the equivalence, i.e., $frev(t)$, is identical to the left-hand side of the equation on the right-hand side of the equivalence of the induction hypothesis. Thus, if we can pull the wave hole out of its nested position and into an equality, we can weak fertilize. This can be achieved by introducing a new existential variable as a placeholder for the wave hole and adding the equality between the new variable and the wave hole. Schematically, this corresponds to applying the rewrite

$$\boxed{\phi(\underline{\psi(x)})} = y :\Rightarrow \boxed{\exists y'.\, \phi(y') = y \wedge \underline{\psi(x) = y'}}, \tag{27}$$

which is not quite a wave rule because it does not entirely preserve the skeleton. The skeleton of the left-hand side is $\psi(x) = y$, that of the right-hand side is $\psi(x) = y'$, where y' is a bound variable. The notion of rippling is currently being extended to allow such rules as wave rules as well (see Bundy and Lombart [11]).

In the example, using the existential variable l', we obtain

$$rrev\left(\boxed{h :: \underline{t}}, l\right) \leftrightarrow \boxed{\exists l'.\, app(l', h :: nil) = l \wedge \underline{frev(t) = l'}}.$$

We can now exploit the induction hypothesis, since l, the variable corresponding to l' in the hypothesis, is universally quantified. We rewrite the conclusion to

$$rrev\left(\boxed{h :: \underline{t}}, l\right) \leftrightarrow \exists l'.\, app(l', h :: nil) = l \wedge rrev(t, l').$$

Very weak fertilization solves a problem that arises in reasoning about relations rather than functions. The need for this technique stems from the fact that we are synthesizing a relational program from a non-tail-recursive function. The recursive case of the reverse function *frev* is non-tail-recursive; that is, its value is defined by a function applied to the result of the recursive call. The flat structure of relations makes such a nesting in the corresponding relation impossible. In the relational case, such nestings can only be expressed by having the corresponding relations share existential variables. Thus, to make progress in the synthesis of a relation, it is necessary to unpack the nesting of functions by introducing existential variables.

6. Implementation and Results

6.1. Implementation

Periwinkle is an adaptation and extension of the *CLAM-Oyster* system [16]. *CLAM* was designed as a planner to plan inductive proofs in type theory. Recently, however, there has been an increased interest in using *CLAM* for other types of proofs, logics, and proof-checkers, and there is an ongoing effort into making *CLAM* less logic-dependent. At the object level, *Oyster* is being replaced by an interactive proof checking shell *Mollusc* [40], which, given a specification of a logic, becomes a proof checker for that logic. Using *Mollusc*, we implemented a proof checker for many-sorted first-order predicate logic with equality and induction, following the Gentzen System $G_=$ from [22] and adding appropriate induction rules.

At the planning level, *CLAM* needed to be adapted to first-order predicate logic with equality. Beyond the syntax-related adaptations, *CLAM* was extended in a number of ways. First, the middle-out reasoning presented here required higher-order pattern unification and a higher-order representation of annotations. Second, we implemented code to handle auxiliary syntheses. Finally, we implemented the new methods and auxiliary code related to these.

6.2. Synthesized Programs

One of our aims was to improve on the results achieved by known synthesis systems, particularly in terms of automation and selection of induction schemes. Three examples from the literature are of special importance: $subset$ from Lau and Prestwich [32], $delete$ from Bundy et al. [13], and qr from Biundo [7]. For the first two examples, the aim was to fully automate the synthesis, which, in the literature, was semi-automatic or interactive. For the last example, the aim was to synthesize a better algorithm by finding a more appropriate induction. *Periwinkle* is able to synthesize good programs for all of the specifications automatically and can thus be considered to have achieved its aims.

The problems taken from the literature can be summarized as follows: the $subset$ example [32]

$$\forall x, y.\, subset(x, y) \leftrightarrow (\forall z.\, member(z, x) \rightarrow member(z, y))$$

posed the problems of rippling using logical wave rules (see Section 5.1) and controlling the rippling to avoid non-termination (see Section 4.4). While the synthesis requires three instances of user interaction in [32], it requires none in *Periwinkle*. The $delete$ example [13]

$$\forall x, y, z.\, delete(x, y, z) \leftrightarrow (\exists k, l.\, fapp(k, l) = y \wedge fapp(k, x :: l) = z)$$

contains nested quantifiers which require unrolling (see Section 5.2). This example, which had not yet been automated at all, is also done automatically by

Periwinkle. A remaining problem with the synthesis of *delete* is that it requires a lemma about the append function *fapp*,

$$\forall k, l.\, fapp(k, l) = nil \leftrightarrow k = nil \wedge l = nil.$$

Finally, the quotient-remainder example *qr* from [7]

$$\forall x, y, q, r.\, qr(x, y, q, r) \leftrightarrow q \times x + r = y \wedge r < x$$

requires middle-out induction (see Section 4) and auxiliary syntheses (see Section 3.5). Biundo's heuristic selects structural induction on y, which leads to an inefficient, unintuitive program. *Periwinkle* chooses a more appropriate induction, namely one where the induction terms for w and y are $s(w)$ and $y + x$, respectively, which leads to a simpler, more efficient program. The synthesis of qr requires three lemmas, the associativity of $plus$, a variant of the cancellation of $plus$, and a lemma on $<$ and $plus$

$$\begin{aligned} \forall x, y, z.\, plus(x, plus(y, z)) &= plus(plus(x, y), z) \\ \forall x, y, z.\, plus(x, y) = plus(x, z) &\leftrightarrow y = z \\ \forall x, y, z.\, (z < x \rightarrow plus(x, y) = z) &\leftrightarrow false. \end{aligned}$$

Table I. Programs synthesized by *Periwinkle*. Planning and plan execution times are for main syntheses in seconds of CPU time on a SPARCstation 10 using Quintus Prolog Release 3.1.4.

Name	Description	Planning Time	Plan Execution Time
add3	Addition relation for three numbers	9.5	5.8
add_length	Combined length of two lists	10.9	5.8
back	see Section 5.2	14.0	5.6
before	An element precedes another in a list	39.1	19.2
between	A number is between two numbers	34.2	4.9
delete	see above	95.0	31.3
even	see Section 4.1	7.2	4.7
insert	Insert an element before another in a list	124.8	37.8
max	see Section 3.5	30.4	10.4
not_member	Non-membership in a list	9.1	3.0
qr	see above	42.1	12.9
rapp	Append relation	8.4	3.7
rplus	Addition relation	6.0	3.6
rrev	see Section 5.3	8.4	4.6
rtimes	Multiplication relation	6.8	5.3
subset	see above	21.2	5.0

Table I gives an overview of other problems solved by *Periwinkle*. All plans were constructed by using middle-out induction. For more detail on the specifications, programs, and proofs, see [29]. The examples fall into three categories:

1. Synthesizing relations from functions. Such relational programs have the advantage that they can be run in various modes. Thus, for instance, an addition relation can also be used for subtraction. Examples programs are $rapp$, $rplus$, $rtimes$, and $rrev$.
2. Synthesizing programs from specifications containing quantified and negated relations. Examples programs are $delete$, $subset$, and max.
3. Synthesizing more efficient programs from executable but less efficient programs. An example program is $between$.

It would not be fair to omit examples that we would have liked to synthesize, but failed to. These include sorting and partitioning lists. The former fails because *Periwinkle* lacks relational rippling (see Section 7.3), the latter because *Periwinkle* is not able to find a required case split.

7. Future Research

7.1. SYNTHESIS

The work in this paper has concentrated on the fully automatic synthesis of recursive logic programs that are partially correct and complete. The system currently tries to find a recursive program, but does not prefer any one type of recursion over another. A useful extension to the system would be to allow the user to specify constraints on the program, for instance by giving the complexity or type of recursion of the program or by requiring that the program be tail-recursive. Allowing the user to determine the recursive structure of the program would shift the challenge to the proof planner from finding appropriate inductions to finding wave rules, since the induction of the proof is determined from the outset.

7.2. MIDDLE-OUT INDUCTION

In middle-out induction, the instantiation of the meta-variables is used as an index into the database of induction schemes. Currently, middle-out induction fails if the ordering determined in the rippling is not among the set of known orderings. However, it may be possible to 'salvage' the induction. For instance, the ordering may be a combination of known orderings. Alternatively, there may be an ordering corresponding to the instantiation of a subset of the meta-variables. In the latter case, the step case would have to be patched by introducing an appropriate case split on the non-induction variable. Ideally, however, the proof planner would not have to rely on a set of known induction schemes, but would use general well-founded induction and prove the well-foundedness of the ordering.

7.3. Rippling

7.3.1. *Search Control*

The control of rippling in middle-out induction poses a considerable challenge. Most important, rippling needs to be controlled so that the search space is kept within reasonable bounds and non-termination is avoided. In Section 4.4, this was achieved by a simple mechanism, i.e., allowing only one speculative step. Although allowing no more than one speculative step ensures termination, it also cuts the search space down too far – there are proofs that require more than one speculative step. An example requiring two speculative steps is

$$\forall w, x, y, z.\, plus(w, x) = plus(y, x) \leftrightarrow plus(w, z) = plus(y, z).$$

With only one speculative step, a blockage occurs that prevents fertilization. What is needed is global control over speculative steps such as speculative rippling, generalization, and lemma conjecturing. Such global control can be provided by the planning critics of Ireland [28], which, given a failed proof planning attempt, analyze it and suggest ways of correcting it. In the case of middle-out induction, the method and its critic would work together as follows: The method allows an initial speculative ripple to begin rippling. Then, it allows only definite rippling and fertilization. If the rippling fails, the critic can analyze the rippling and suggest the appropriate measure – another speculative ripple, a lemma, or a generalization.

7.3.2. *Rippling with Relations*

The class of programs we can synthesize so far is limited mainly because rippling as presented by Bundy et al. [14] is based on nested functions. The idea of a wave rule is precisely that it moves constructors or functions up from a nested position. To extend the class of logic programs we can synthesize, the notion of rippling needs to be extended to relations. Relations cannot be nested like functions, but a similar effect is achieved by existentially quantified variables that are arguments of more than one literal. Take, for instance, the step cases of standard functional and relational definitions of list reversal:

$$\forall h, t.\, frev(h :: t) \;=\; append(frev(t), h :: nil) \tag{28}$$
$$\forall h, t.\, rrev(h :: t, l) \;\leftrightarrow\; \exists l'.\, rrev(t, l') \wedge append(l', h :: nil, l). \tag{29}$$

While (28) gives rise to a wave rule

$$frev\left(\boxed{H :: \underline{T}}\right) :\Rightarrow \boxed{append(\underline{frev(T)}, H :: nil)},$$

(29) currently does not. The rule

$$rrev\left(\boxed{H :: \underline{T}}, L\right) :\Rightarrow \boxed{\exists l'.\, \underline{rrev(T, l')} \wedge append(l', H :: nil, L)}$$

is not a wave rule in the traditional sense because the skeleton of the left-hand side, $rrev(T, L)$, and the skeleton of the right-hand side, $rrev(T, l')$, are not identical. This 'almost' wave rule is very similar to the schematic rewrite (27) in Section 5 in that they both preserve the skeleton except for the names of bound variables. Such rules are clearly within the spirit of rippling, and the notion of rippling is being extended to cope with them (see Bundy and Lombart [11]). Relational rippling would enable us to synthesize a large number of programs that are currently beyond the capabilities of the system.

8. Conclusions

In this paper, we have investigated the application of proof planning to the automatic synthesis of logic programs via inductive proofs. The work developed out of existing work in middle-out reasoning [12] and in proofs-as-programs for logic program synthesis [13]. The main goals were to synthesize relational programs and to automate the synthesis process. Our work makes four principal contributions:

1. We have applied middle-out reasoning to new applications and have shown how to extend it to overcome related search problems. For example, we show how search arising in higher-order unification can be restricted by the use of higher-order patterns.
2. We have shown that middle-out reasoning is a mechanism through which proof planning can be used to synthesize logic programs fully automatically.
3. We have provided evidence that middle-out reasoning provides an elegant and effective mechanism to select induction schemes. Middle-out induction has made the class of synthesis theorems, for which existing techniques often fail, amenable to automatic proof.
4. We have provided new techniques for unblocking rippling when the blockage is caused by missing wave rules on propositional connectives or by nested quantifiers.

Although we are still far away from automatically synthesizing complex programs from formal specifications, we have made progress towards that goal by automating some of the steps that are traditionally considered eureka steps, and are therefore normally left to the user. Middle-out reasoning has played a crucial role in this success by providing a elegant framework for speculation, and we believe that its potential is far from exhausted.

Notes

[1] In logic programming, partial correctness means that the computed relation is a subset of the specified relation, completeness that the specified relation is a subset of the computed relation [18, 17].

[2] We use the variation here because *Periwinkle* fails on the original, but succeeds for the variation. Though it does find the appropriate induction in the original, i.e., two-step induction on x, it fails on the auxiliary synthesis in the second base case

$$\forall x.\, auxeven(x) \leftrightarrow (\exists y.\, y \times s(s(0)) = s(0)),$$

since it is not yet able to simplify $\exists y.\, y \times s(s(0)) = s(0)$ to false.

[3] Here, we would need to generalize before doing induction. In the ordering of methods, however, induction comes before generalization, and the depth-first planner will select induction. This particular example can be solved by a best-first planner with an evaluation function to determine whether induction or generalization should be preferred (see Manning [35]).

Acknowledgements

We thank Sophie Renault for her careful reading of this paper.

References

1. Aubin, R.: Mechanizing structural induction, Ph.D. thesis, University of Edinburgh, 1976.
2. Basin, D., Bundy, A., Kraan, I., and Matthews, S.: A framework for program development based on schematic proof, in *Proc. 7th Int. Workshop on Software Specification and Design (IWSSD-93)*, 1993. Also available as Max-Planck-Institut für Informatik Report MPI-I-93-231 and Edinburgh DAI Research Report 654.
3. Basin, D. and Walsh, T.: A calculus for and termination of rippling, *J. Automated Reasoning* **16**(1–2) (1996), 145–178.
4. Bibel, W.: Syntax-directed, semantics-supported program synthesis, *Artificial Intelligence* **14** (1980), 243–261.
5. Bibel, W. and Hörnig, K. M.: LOPS – a system based on a strategical approach to program synthesis, in A. Biermann, G. Guiho, and Y. Kodratoff (eds), *Automatic Program Construction Techniques*, Macmillan, 1984, pp. 69–90.
6. Biundo, S.: Automated synthesis of recursive algorithms as a theorem proving tool, in Y. Kodratoff (ed.), *8th Eur. Conf. Artificial Intelligence*, Pitman, 1988, pp. 553–558.
7. Biundo, S.: Automatische Synthese rekursiver Programme als Beweisverfahren, Ph.D. thesis, Universität Karlsruhe, 1989.
8. Biundo, S., Hummel, B., Hutter, D., and Walther, C.: The Karlsruhe induction theorem proving system, in J. Siekmann (ed.), *8th Conf. Automated Deduction*, Lecture Notes in Computer Science 230, Springer-Verlag, 1986, pp. 672–674.
9. Boyer, R. S. and Moore, J S.: *A Computational Logic Handbook*, Academic Press, 1988. Perspectives in Computing, Vol. 23.
10. Bundy, A.: The use of explicit plans to guide inductive proofs, in R. Lusk and R. Overbeek (eds), *9th Conf. Automated Deduction*, Springer-Verlag, 1988, pp. 111–120. Longer version available from Edinburgh as DAI Research Paper 349.
11. Bundy, A. and Lombart, V.: Relational rippling: A general approach, in C. Mellish (ed.), *Proc. 14th Int. Joint Conf. Artificial Intelligence*, Morgan Kaufmann, 1995, pp. 175–181.
12. Bundy, A., Smaill, A., and Hesketh, J.: Turning eureka steps into calculations in automatic program synthesis, in S. L. H. Clarke (ed.), *Proc. UK IT 90*, 1990, pp. 221–226.
13. Bundy, A., Smaill, A., and Wiggins, G. A.: The synthesis of logic programs from inductive proofs, in J. Lloyd (ed.), *Computational Logic*, Springer-Verlag, 1990, pp. 135–149. Esprit Basic Research Series. Also available from Edinburgh as DAI Research Paper 501.
14. Bundy, A., Stevens, A., van Harmelen, F., Ireland, A., and Smaill, A.: Rippling: A heuristic for guiding inductive proofs, *Artificial Intelligence* **62** (1993), 185–253. Also available from Edinburgh as DAI Research Paper 567.
15. Bundy, A., van Harmelen, F., Hesketh, J., Smaill, A., and Stevens, A.: A rational reconstruction and extension of recursion analysis, in N. S. Sridharan (ed.), *Proc. 11th Int. Joint Conf. Artificial

Intelligence, Morgan Kaufmann, 1989, pp. 359–365. Also available from Edinburgh as DAI Research Paper 419.
16. Bundy, A., van Harmelen, F., Horn, C., and Smaill, A.: The Oyster-Clam system, in M. E. Stickel (ed.), *10th Int. Conf. Automated Deduction*, Lecture Notes in Artificial Intelligence 449, Springer-Verlag, 1990, pp. 647–648. Also available from Edinburgh as DAI Research Paper 506.
17. Clark, K. L.: Predicate Logic as a Computational Mechanism, Technical Report 79–59, Dept. of Computing, Imperial College, 1979.
18. Clark, K. L. and Tärnlund, S.-Å.: A first order theory of data and programs, in B. Gilchrist (ed.), *Information Processing* IFIP, 1977, pp. 939–944.
19. Deville, Y.: *Logic Programming. Systematic Program Development*, Int. Series in Logic Programming, Addison-Wesley, 1990.
20. Deville, Y. and Lau, K. K.: Logic program synthesis, *J. Logic Programming* **19/20** (May/July 1994), 321–350.
21. Fribourg, L.: Extracting logic programs from proofs that use extended Prolog execution and induction, in *Proc. 8th Int. Conf. Logic Programming*, MIT Press, June 1990, pp. 685–699.
22. Gallier, J.: *Logic for Computer Science*, Harper & Row, New York, 1986.
23. Gordon, M. J., Milner, A. J., and Wadsworth, C. P.: *Edinburgh LCF – A Mechanised Logic of Computation*, Lecture Notes in Computer Science 78, Springer-Verlag, 1979.
24. Hesketh, J. T.: Using middle-out reasoning to guide inductive theorem proving, Ph.D. thesis, University of Edinburgh, 1991.
25. Huet, G.: A unification algorithm for lambda calculus, *Theoretical Computer Science* **1** (1975), 27–57.
26. Hutter, D.: Synthesis of induction ordering for existence proofs, in *Proc. 12th Conf. Automated Deduction*, 1994.
27. Ireland, A.: The use of planning critics in mechanizing inductive proofs, in A. Voronkov (ed.), *Int. Conf. Logic Programming and Automated Reasoning (LPAR 92), St. Petersburg*, Lecture Notes in Artificial Intelligence 624, Springer-Verlag, 1992, pp. 178–189. Also available from Edinburgh as DAI Research Paper 592.
28. Ireland, A. and Bundy, A.: Productive use of failure in inductive proof, *J. Automated Reasoning* **16**(1–2) (1996), 79–111.
29. Kraan, I.: Proof planning for logic program synthesis, Ph.D. thesis, University of Edinburgh, 1994.
30. Kraan, I., Basin, D., and Bundy, A.: Logic program synthesis via proof planning, in K. K. Lau and T. Clement (eds), *Logic Program Synthesis and Transformation*, Springer-Verlag, 1993, pp. 1–14. Also available as Max-Planck-Institut für Informatik Report MPI-I-92-244 and Edinburgh DAI Research Report 603.
31. Kraan, I., Basin, D., and Bundy, A.: Middle-out reasoning for logic program synthesis, in D. S. Warren (ed.), *Proc. 10th Int. Conf. Logic Programming*, MIT Press, 1993. Also available as Max-Planck-Institut für Informatik Report MPI-I-93-214 and Edinburgh DAI Research Report 638.
32. Lau, K. K. and Prestwich, S. D.: Synthesis of Recursive Logic Procedures by Top-Down Folding, Technical Report UMCS-88-2-1, Dept. of Computer Science, University of Manchester, February 1988.
33. Lau, K. K. and Prestwich, S. D.: Top-down synthesis of recursive logic procedures from first-order logic specifications, in D. H. D. Warren and P. Szeredi (eds), *Proc. 7th Int. Conf. Logic Programming*, MIT Press, 1990, pp. 667–684.
34. Lloyd, J. W.: *Foundations of Logic Programming, Symbolic Computation*, Springer-Verlag, 1987, second, extended edition.
35. Manning, A.: Representing preference in proof plans, Master's thesis, Edinburgh, 1992.
36. Martin-Löf, P.: Constructive mathematics and computer programming, in *6th Int. Congr. for Logic, Methodology and Philosophy of Science*, North Holland, Amsterdam, 1982, pp. 153–175.

37. Miller, D.: A logic programming language with lambda abstraction, function variables and simple unification, in *Extensions of Logic Programming*, Lecture Notes in Artificial Intelligence 475, Springer-Verlag, 1991.
38. Nipkow, T.: Higher-order critical pairs, in *Proc. 6th IEEE Symp. Logic in Computer Science*, 1991, pp. 342–349.
39. Protzen, M.: Lazy generation of induction hypotheses, in A. Bundy (ed.), *Proc. 12th Conf. on Automated Deduction*, Lecture Notes in Artificial Intelligence 814, Springer-Verlag, 1994.
40. Richards, B. L., Kraan, I., Smaill, A., and Wiggins, G. A.: Mollusc: A general proof development shell for sequent-based logics, in A. Bundy (ed.), *Proc. 12th Conf. Automated Deduction*, Lecture Notes in Artificial Intelligence 814, Springer-Verlag, 1994, pp. 826–830. Also available from Edinburgh as DAI Research Paper 723.
41. Stevens, A.: A rational reconstruction of Boyer and Moore's technique for constructing induction formulas, in Y. Kodratoff (ed.), *Proc. European Conference on Artificial Intelligence (ECAI-88)*, 1988, pp. 565–570. Also available from Edinburgh as DAI Research Paper 360.
42. Wiggins, G. A.: Synthesis and transformation of logic programs in the Whelk proof development system, in K. R. Apt (ed.), *Proc. Joint Int. Conf. Symp. Logic Programming (JICSLP-92)*, MIT Press, Cambridge, MA, 1992, pp. 351–368.

Journal of Automated Reasoning **16:** 147–180, 1996.

A Calculus for and Termination of Rippling

DAVID A. BASIN*
Max-Planck-Institut für Informatik, Saarbrücken, Germany
e-mail: basin@mpi-sb.mpg.de

TOBY WALSH**
Mechanized Reasoning Group, DIST, Genova and IRST, Trento, Italy
e-mail: toby@irst.it

Abstract. Rippling is a type of rewriting developed for inductive theorem proving that uses annotations to direct search. Rippling has many desirable properties: for example, it is highly goal directed, usually involves little search, and always terminates. In this paper we give a new and more general formalization of rippling. We introduce a simple calculus for rewriting annotated terms, close in spirit to first-order rewriting, and prove that it has the formal properties desired of rippling. Next we develop criteria for proving the termination of such annotated rewriting, and introduce orders on annotated terms that lead to termination. In addition, we show how to make rippling more flexible by adapting the termination orders to the problem domain. Our work has practical as well as theoretical advantages: it has led to a very simple implementation of rippling that has been integrated in the Edinburgh CLAM system.

Key words: Mathematical induction, inductive theorem proving, term rewriting.

1. Introduction

Rippling is a form of rewriting developed by Bundy et al. [6, 8] that uses annotations to restrict rewriting and to guide the derivation towards a particular goal. Rippling applies naturally in inductive theorem proving where the induction conclusion typically differs from the induction hypothesis by the addition of some constructors or destructors. These differences are marked by annotations. Rippling uses annotated rewrite rules, called *wave-rules*, to move these marked differences; when successful, the differences are either removed completely or moved to positions like the top of the term that enable the use of the induction hypothesis.

Rippling has many attractive properties. It is highly goal directed, manipulating just the differences between the induction conclusion and hypothesis while

* Funded by the German Ministry for Research and Technology under grant ITS 9102.
** Supported by a Human Capital and Mobility Research Fellowship from the European Commission. Both authors thank members of the Edinburgh Mathematical Reasoning Group, as well as Alan Bundy, Leo Bachmair, Dieter Hutter, and Michael Rusinowitch, for their comments on previous drafts. Additional support was also received from the MInd grant EC-US 019-76094.

leaving their common structure preserved; this is in contrast to rewriting based on normalization, which is used in other inductive theorem provers such as Nqthm [4]. Rippling also involves little search, since annotations severely restrict rewriting. Indeed, the use of annotation places such strong restrictions on the search space that it is often possible to analyze failed rippling proofs and to suggest missing lemmas or generalizations [15].

In this paper we give a new account of rippling and its properties that is both substantially simpler and more general than previous accounts. Conceptually, the starting point of our work is the formal presentation of rippling given by Bundy et al. in [6]. There rippling is presented as first-order rewriting restricted by some simple preconditions that ensure that annotations in the subterm being rewritten match annotations in the applied wave-rule. We show that this is inadequate. Taken literally, it leads to an implementation of rippling too restrictive to carry out the derivations given in [6] (see Section 8). When the restrictions are loosened (as they were in the implementation of the CLAM system [7]), other problems arise, such as improperly annotated terms. We give a simple calculus for rippling that does not suffer from these problems.

We also simplify, improve, and generalize the specification of wave-rules and their associated termination orderings. Wave-rules have previously been presented using complex schematic definitions that combine the properties of structure preservation and the reduction of a well-founded measure (see Section 8). Our definition of wave-rules separates these two concerns and their demonstration. We present measures that, despite their simplicity, admit strictly more wave-rules than the considerably more complex specification given in [6].

Another contribution of our work is to provide new termination orderings that extend the power of rippling. Although rippling was designed primarily to prove inductive theorems, it has recently been applied to other problem domains. For example, it has been used to sum series [16], to prove limit theorems [17], and to perform normalization [1]. In rippling, as in conventional rewriting, the termination ordering can be made domain-dependent. We illustrate this idea by two new orderings.

A practical contribution of our work is that it greatly simplifies implementation; our calculus leads to an implementation of rippling in the spirit of standard first-order rewriting. Moreover, systems implementing rippling require a procedure which annotates rewrite rules (here called a *wave-rule parser*) and our work eases the construction of such a routine. We report on our implementation of these routines, which has been integrated into the CLAM system.

The remainder of this paper is organized as follows. Section 2 provides a background on rippling, in particular on annotation and wave-rules. Section 3 formalizes properties of rippling and shows why first-order rewriting cannot directly satisfy these properties. Section 4 presents a new calculus that does satisfy these properties. Next, Section 5 introduces reduction orders under which rippling will terminate. Section 6 describes how our calculus can be implemented and how unannotated rewrite rules can be annotated and oriented to give wave-rules. Section 7 presents two new orders which extend the power of rippling. Finally, in Section 8 we survey related work.

2. Annotation and Wave-Rules

In this section, we give a brief overview of rippling and introduce notation and terminology. Reference [6] should be consulted for additional motivation and examples.

Rippling arose out of an analysis of inductive proofs and of the heuristics embedded within the Nqthm theorem prover [4]. As a simple example, suppose we wish to prove $P(x)$ for all natural numbers, x. We assume $P(n)$ and attempt to show $P(s(n))$. The hypothesis and the conclusion are identical except for the successor function $s(\cdot)$ applied to the induction variable n. Rippling marks this difference by the annotation, $P(\boxed{s(\underline{n})}^{\uparrow})$. The annotation in the induction conclusion, given by the box, arrow, and underlining, marks the differences with the induction hypothesis. Deleting the arrow and everything in the box that is not underlined gives the *skeleton*, $P(n)$; this is identical to the induction hypothesis and is preserved during rippling. By comparison, simply removing annotations gives the *erasure*, $P(s(n))$. The boxed but not underlined term parts are *wave-fronts*; these are moved and transformed (and possibly deleted) by rippling. The underlined parts are *wave-holes*; they represent terms in the wave-front that we wish to leave unchanged. Wave-fronts are marked with arrows indicating if rippling should try to move the wave-front up through the skeleton term tree or down towards the leaves. Wave-fronts with an up arrow are called *outward* directed, while those with a down arrow are *inward* directed.

A wave-front can be viewed as a context; that is, it is a term with one, or more, proper subterms deleted. Schematically, an outward directed wave-front is of the form $\boxed{\xi(\underline{\mu_1}, \ldots, \underline{\mu_n})}^{\uparrow}$ where $n > 0$ and the μ_i may be similarly annotated; when $n = 1$ we call the wave-front *simply annotated* and when $n > 1$ we call it *multi-hole annotated*. A term is said to be simply annotated when all its wave-fronts are simply annotated, and is multi-hole annotated otherwise.

To formalize rippling, we extend the signature of the original theory with three new unary function symbols: $wfout$, $wfin$, and wh (representing outward and inward directed wave-fronts and wave-holes, respectively). The requirement that a context has at least one hole and that terms in these holes may be further annotated can be formally captured by defining the set of well-annotated terms with respect to a set of unannotated (first-order) terms *unats*.

DEFINITION 1. *Well-annotated terms* (or *wats*) are the smallest set such that

1. t is a wat for all unannotated terms;

2. $wfout(f(t_1,\ldots,t_n))$ is a wat iff for at least some i, $t_i = wh(s_i)$ and for each i where $t_i = wh(s_i)$, s_i is a wat and for each i where $t_i \neq wh(s_i)$, t_i is an unannotated term;
3. $wfin(f(t_1,\ldots,t_n))$ is a wat under the same conditions in case (2).
4. $f(t_1,\ldots,t_n)$ is a wat, where f is not $wfout$, $wfin$, or wh iff each t_i is a wat.

For example, $wfout(s(wh(x))) \times wfout(s(wh(y)))$ represents the well-annotated term $\boxed{s(\underline{x})}^{\uparrow} \times \boxed{s(\underline{y})}^{\uparrow}$. To simplify the presentation of annotated terms, we shall still use boxes, arrows, and holes. However, this is *just* syntactic sugar for $wfout$, $wfin$, and wh. Note that in our formalization of annotation, wave-holes occur as immediate subterms of the function symbol in the wave-front. As additional syntactic sugar, we may merge adjacent wave-fronts and wave-holes when displaying annotated terms. For example, we will display the term $\boxed{s(\underline{\boxed{s(\underline{x})}^{\uparrow}})}^{\uparrow}$ as the annotated term $\boxed{s(s(\underline{x}))}^{\uparrow}$. Insisting that wave-fronts are 'maximally split' both simplifies the presentation of rippling that follows and leads to a simpler implementation since wave-fronts do not need to be dynamically split or merged during rippling, as in [6].

We define the skeleton of an annotated term to be the set of unannotated terms formed by deleting function symbols and variables within wave-fronts that are not within wave-holes.

DEFINITION 2. The *skeleton function skel*: *wats* $\rightarrow \mathcal{P}$(*unats*) is defined by
1. $skel(x) = \{x\}$ for all variables x;
2. $skel(\boxed{f(t_1,\ldots,t_n)}^{\uparrow}) = \{s \mid \exists i.t_i = \underline{t_i'} \wedge s \in skel(t_i')\}$;
3. $skel(\boxed{f(t_1,\ldots,t_n)}^{\downarrow}) = \{s \mid \exists i.t_i = \underline{t_i'} \wedge s \in skel(t_i')\}$;
4. $skel(f(t_1,\ldots,t_n)) = \{f(s_1,\ldots,s_n) \mid \forall i.s_i \in skel(t_i)\}$.

For example, the skeleton of the annotated term

$$s(\boxed{\underline{\boxed{s(\underline{a})}^{\uparrow}} + (\underline{b \times c})}^{\uparrow}) \tag{1}$$

is the set $\{s(a), s(b \times c)\}$. Note that the skeleton of a simply annotated term is a singleton set; in this case, we refer to the member as *the skeleton* of the term.

By erasing annotation, we construct the corresponding unannotated term.

DEFINITION 3. The *erasure function erase*: *wats* $\rightarrow$ *unats* is defined by
1. $erase(x) = x$ for all variables x;

2. $erase(\boxed{f(t_1,\ldots,t_n)}^{\uparrow}) = f(s_1,\ldots,s_n)$ where if $t_i = \underline{t'_i}$, then $s_i = erase(t'_i)$, else $s_i = t_i$;
3. $erase(\boxed{f(t_1,\ldots,t_n)}^{\downarrow}) = f(s_1,\ldots,s_n)$ where if $t_i = \underline{t'_i}$, then $s_i = erase(t'_i)$, else $s_i = t_i$;
4. $erase(f(t_1,\ldots,t_n)) = f(s_1,\ldots,s_n)$ where $s_i = erase(t_i)$.

For example, the erasure of (1) is $s(s(a) + (b \times c))$.

2.1. WAVE-RULES AND RIPPLING

Wave-rules are formally defined in Subsection 4.4. Informally, they are rewrite rules between annotated terms that are skeleton preserving and measure decreasing under an appropriate ordering on annotated terms. This definition is simpler and more general than the one given in [6], where these two requirements were combined in the syntactic specification of a wave-rule.[1]

Skeleton preservation in the simply annotated case means that both the LHS (left-hand side) and RHS (right-hand side) of the wave-rule have an identical skeleton. In the multihole case we demand that *some* of the skeletons on the LHS are preserved on the RHS and no new skeletons are introduced, i.e., $skel(LHS) \supseteq skel(RHS)$. *Measure reduction* and orderings for annotated terms will be described in considerable detail later. The intuition is that the position and orientation of wave-fronts define a measure. Rippling makes well-founded progress by moving annotations to decrease this measure, for example, moving outwards-directed wave-fronts upwards in the rewritten term.

To illustrate the different types of wave-rules that our definition captures and to help motivate the definition of termination orders on annotated terms, we give some examples of wave-rules used by rippling. To begin with, most recursive function definitions and all primitive recursive function definitions can be annotated as wave-rules. For example, the recursive definition of times gives the wave-rule

$$\boxed{s(\underline{U})}^{\uparrow} \times V \rightarrow \boxed{\underline{(U \times V)} + V}^{\uparrow}. \tag{2}$$

Lemmas also can often be annotated as wave-rules, for example, algebraic laws like associativity, distributivity, and cancellation ($<>$ is infix append on lists).

$$(\boxed{\underline{U} + V}^{\uparrow}) \times W \rightarrow \boxed{\underline{U \times W} + V \times W}^{\uparrow} \tag{3}$$

$$\boxed{(U <> \underline{V})}^{\uparrow} <> W \rightarrow \boxed{U <> \underline{(V <> W)}}^{\uparrow} \tag{4}$$

$$U <> (\boxed{\underline{V} <> W}^{\uparrow}) \rightarrow \boxed{\underline{(U <> V)} <> W}^{\uparrow} \tag{5}$$

$$\boxed{\underline{U} + \underline{V}}^{\uparrow} = \boxed{\underline{W} + \underline{Z}}^{\uparrow} \rightarrow \boxed{\underline{U = W} \wedge \underline{V = Z}}^{\uparrow} \tag{6}$$

(4) and (5) demonstrate that rules like associativity can be wave-rules in both directions; the precondition on rippling that annotations in wave-rules match those in the rewritten term prevent looping. (6) is an example of a wave-rule with multihole annotation. Note that $\rightarrow$ indicates rewriting, not implication; in fact, implication in (6) holds only in the reverse direction. This is sensible because in inductive theorem proving, rippling reasons backwards from the induction conclusion to the induction hypothesis and rewrite rules based on implication as opposed to equivalence may be used; see [6] for further discussion on how such rewrite rules may be used in forward directed theorem proving.

Wave-rules can also ripple wave-fronts inwards, or change the orientation of wave-fronts from out to in (but not in to out).

$$\boxed{\underline{(U <> V)} <> W}^{\downarrow} \rightarrow U <> (\boxed{\underline{V} <> W}^{\downarrow}) \tag{7}$$

$$(\boxed{\underline{U} <> V}^{\uparrow}) <> W \rightarrow U <> (\boxed{V <> \underline{W}}^{\downarrow}) \tag{8}$$

A simple example of a proof guided by rippling is the proof of the associativity of multiplication

$$(x \times y) \times z = x \times (y \times z). \tag{9}$$

The proof uses structural induction on x. In the step-case, (9) is the induction hypothesis and the induction conclusion is

$$(\boxed{s(\underline{x})}^{\uparrow} \times y) \times z = \boxed{s(\underline{x})}^{\uparrow} \times (y \times z).$$

The wave-fronts in the induction conclusion mark the differences with the induction hypothesis. Rippling on both sides of the induction conclusion using (2) yields

$$(\boxed{\underline{x \times y} + y}^{\uparrow}) \times z = \boxed{\underline{(x \times (y \times z))} + y \times z}^{\uparrow}.$$

Rippling with (3) on the left-hand side then gives

$$\boxed{\underline{(x \times y) \times z} + y \times z}^{\uparrow} = \boxed{\underline{x \times (y \times z)} + y \times z}^{\uparrow}.$$

As expected, the skeleton in each step of our proof is the induction hypothesis. At the end of this rewriting, the wave-fronts are at the root of each term. We can therefore complete the proof by simplifying with the induction hypothesis.

Rippling outward directed wave-fronts up towards the root of terms, like in the previous example, is called *rippling-out*. Wave-rules can also ripple wave-fronts downwards towards the position of universally quantified variables in the induction hypothesis. Such positions are called *sinks* because wave-fronts can be 'absorbed' there: when we appeal to the induction hypothesis, universally quantified variables will be matched with the content of the sinks; an example of this is provided in Subsection 7.1. Rippling downward-directed wave-fronts towards the leaves of the term tree using rules like (7) is called *rippling-in*.

3. Properties of Rippling

We now formalize the properties desired of rippling which motivate our calculus.

Well-formedness: if s is a wat, and s ripples to t, then t is also a wat;

Skeleton preservation: if s ripples to t, then $skel(t) \subseteq skel(s)$;

Correctness: if s ripples to t, then $erase(s)$ rewrites to $erase(t)$ in the original (unannotated) theory;

Termination: rippling terminates.

The first property means that terms manipulated by rippling stay well formed. Hence we can always compute their skeleton and erasure. Skeleton preservation ensures that rippling directs the derivation towards (at least) one of the induction hypotheses. Correctness guarantees that we can perform the corresponding derivation in the underlying object-level theory; annotation merely guides search. Finally, termination is important for practical considerations; it means we can try other possibilities (e.g., other inductions) when derivations fails.

Unfortunately, rippling implemented directly by first-order rewriting in an annotated theory fails to achieve these properties. We demonstrate this negative result by example. Consider using first-order rewriting to apply the wave-rule corresponding to the recursive definition of multiplication given in (2) to the term $\boxed{s(\underline{x})}^{\uparrow} \times \boxed{s(\underline{y})}^{\uparrow}$. This is a wat but rewrites to $\boxed{\underline{x \times \boxed{s(\underline{y})}^{\uparrow}} + \boxed{s(\underline{y})}^{\uparrow}}^{\uparrow}$, which is not a wat because the second argument of plus contains a wave-front (a box) directly inside another without an intermediate wave-hole. The reader may find it easier to carry out the rewrite step and verify this by removing the syntactic sugar and representing the boxes and holes explicitly with $wfout$ and wh.

Termination also fails using first-order rewriting directly. Consider, for example, the equation $h(f(U, s(V))) = s(h(f(s(U), V)))$. Under the width measure given in Subsection 5.2, this is a wave-rule in the left-to-right direction annotated as

$$h(\boxed{f(U, s(\underline{V}))}^{\uparrow}) \rightarrow \boxed{s(\underline{h(\boxed{f(s(U), \underline{V})}^{\uparrow})})}^{\uparrow}.$$

It is also a wave-rule in the reverse direction with the annotation

$$\boxed{s(\underline{h(\boxed{f(s(\underline{U}), V)}^{\uparrow})})}^{\uparrow} \rightarrow h(\boxed{f(\underline{U}, s(V))}^{\uparrow}).$$

However, these wave-rules together lead to cycling as follows:

$$h(\boxed{f(\underline{a}, s(\underline{a}))}^{\uparrow}) \mapsto \boxed{s(\underline{h(\boxed{f(s(\underline{a}), \underline{a})}^{\uparrow})})}^{\uparrow} \mapsto h(\boxed{f(\underline{a}, s(\underline{a}))}^{\uparrow}) \mapsto \cdots .$$

Note that unlike the multiplication example, all terms in this derivation are well annotated and share the same skeleton. With two equations, we can construct looping derivations.

Both the problems of improperly annotated terms and non-termination arise when an annotated term replaces an unannotated term in a wave-front. This observation motivates our definition of a calculus for rippling. We introduce a new notion of term replacement that erases annotation when replacing terms in a wave-front. The redefinition of term replacement naturally gives rise to a new notion of substitution, and thus matching. By means of these simple modifications, we develop a calculus for rewriting annotated terms that is guaranteed to preserve well-formedness of annotation, skeletons, and correctness with respect to the underlying theory. In addition, this calculus allows us to design simple termination measures that are stable and monotonic. For unannotated terms and rewrite rules, our calculus performs conventional rewriting.

4. A Calculus for Annotated Rewriting

Notational Convention: To simplify notation in proofs, in this section we assume that the arguments of an annotated term like $\boxed{f(t_1, \ldots, t_n)}^{\uparrow}$ may be partitioned so that the first j arguments are headed by wave-holes, i.e., of the form $t_i = \underline{t_i'}$ for $i \in \{1, \ldots, j\}$, and the last $n - j$ are unannotated. Hence the term may be written as $\boxed{f(\underline{t_1}, \ldots, \underline{t_j}, t_{j+1}, \ldots, t_n)}^{\uparrow}$, or even $\boxed{f(\underline{t_1}, \ldots, t_n)}^{\uparrow}$. This is without loss of generality, as the proofs below do not depend on the order of wave-holes.

4.1. Ground Rewriting

We first consider rewriting using ground rewrite rules. As is typical (e.g., see [11, Section II]), we distinguish between two kinds of variables: those in rewrite rules and those in terms. We treat the later kind, 'term variables', as constants.

We begin by redefining subterm replacement. Let $s[l]$ represent a *wat* with a distinguished subterm l that is also a *wat*. Let $rep(s[l], r)$ denote subterm replacement of r for the distinguished occurrence of l in the term s; *rep* is defined identically to the usual subterm replacement of l by r in s except that if l occurs within a wave-front, any annotations on r are erased before replacement. For example, replacing a in $\boxed{\underline{b} + s(a)}^{\uparrow}$ by $\boxed{s(\underline{a})}^{\uparrow}$ gives $\boxed{\underline{b} + s(s(a))}^{\uparrow}$, but replacing

b by $\boxed{s(\underline{b})}^{\uparrow}$ gives $\boxed{\underline{\underline{\boxed{s(\underline{b})}^{\uparrow}}} + s(a)}^{\uparrow}$. From now on, we will perform all term replacement (including that occurring during substitution) using this function.

In rewriting terms, we will use just *proper* rewrite rules.

DEFINITION 4. $l \to r$ is a *proper rewrite rule* when $erase(l) \to erase(r)$ is a rewrite rule, l and r are *wats*, and $skel(r) \subseteq skel(l)$.

The last two requirements are needed for well-formedness and skeleton preservation. Note also that the requirement that $erase(l) \to erase(r)$ is a rewrite rule means that $Vars(r) \subseteq Vars(l)$.

Ground rewriting consists of rewriting using proper rewrite rules that contain no (non-term) variables. Let R be a set of rewrite rules that are (for now) ground. If $s[l]$ is a *wat*, then ground rewriting with a rule $l \to r$ in R yields $rep(s[l], r)$; we use $s[l] \mapsto_R s[r]$ to denote such *ground rippling*. In what follows we will assume a particular rewrite rule set R, and drop subscripted references, e.g., writing simply $s[l] \mapsto s[r]$.

Ground rippling preserves the well-formedness of annotated terms, preserves skeletons, and corresponds to an annotated version of rewriting in the underlying (unannotated) theory.

THEOREM 1. *If s is a* wat, *$l \to r$ a proper rewrite rule between ground* wats *l and r and $s[l] \mapsto s[r]$, then*

1. *$s[r]$ is a* wat,
2. *$skel(s[r]) \subseteq skel(s[l])$,*
3. *$erase(s[l]) \to erase(s[r])$.*

Proof. (*sketch*) The proof follows by structural induction on s. The only non-trivial case occurs when s is headed by a wave-front, e.g.,

$$s = \boxed{f(\underline{s_1}, \ldots, \underline{s_j}, s_{j+1}, \ldots, s_n)}^{\uparrow},$$

and l is *strict* subterm of one of the s_i (the case for inward directed wave-fronts is analogous). There are two cases depending on if $i \leqslant j$. In the first case, s_i is a *wat*; by the induction hypothesis, $s_i[r]$ is a *wat*. Thus, $s[r]$ is a *wat*. Also, by the induction hypothesis $skel(s_i[r]) \subseteq skel(s_i[l])$. As no other subterm is changed, the union of their skeletons is unchanged. Hence $skel(s[r]) \subseteq skel(s[l])$. Finally, by the induction hypothesis, $erase(s_i[l]) \to erase(s_i[r])$. Again, as all the other subterms are unchanged, their erasures stay the same. Thus, $erase(s[l]) \to erase(s[r])$. In the second case, s_i is an unannotated term within the wave-front. Thus, when we substitute r for l we will erase annotations on r. Hence, $s_i[r]$ is unannotated, and $s[r]$ is a *wat*. From the definition of the skeleton it follows that term replacement in wave-fronts has no effect on the skeleton, so $skel(s[l]) = skel(s) = skel(s[r])$. Finally, s_i is unannotated and $l \to r$ is a proper equation, $erase(s_i[l]) \to erase(s_i[r])$, and thus $erase(s[l]) \to erase(s[r])$. □

By Theorem 1 it follows by induction on the number of rewrite steps that the reflexive transitive closure of $\mapsto$ on ground *wats* also preserves well-formedness, skeletons, and correctness with respect to the theory.

As a simple example, let $\boxed{h(\underline{a})}^{\uparrow} \to a$ be a proper rewrite rule. We can apply this rule only to the first subterm of the wat

$$\boxed{f(\underline{\boxed{h(\underline{a})}^{\uparrow}}, h(a))}^{\uparrow} \qquad (10)$$

and this results in $\boxed{f(\underline{a}, h(a))}^{\uparrow}$. Alternatively, we could apply the proper rewrite rule $a \to \boxed{h(\underline{a})}^{\uparrow}$ to both occurrences of a in (10), resulting in

$$\boxed{f(\underline{\boxed{h(\underline{\boxed{h(\underline{a})}^{\uparrow}})}^{\uparrow}}, h(h(a)))}^{\uparrow} . \qquad (11)$$

Note that annotation was erased when substituting $\boxed{h(\underline{a})}^{\uparrow}$ for the second occurrence of a. We can apply this rewrite rule again to both occurrences of a in (11). While rippling with proper rewrite rules is structure preserving (the skeleton of the rewritten term is always the same as the skeleton of the original term), this example shows that it is not necessarily terminating; for termination we need further restrictions, which will be introduced later.

4.2. Annotated Matching

When rewrite rules contain (non-term) variables, they must be applied by using matching. Since substitution depends on subterm replacement, our new definition of subterm replacement gives rise to a new kind of substitution; during substitution, terms replacing variables in wave-fronts are erased of annotation. This in turn gives rise to a new kind of matching. We represent a *well-annotated substitution* (or *was*) by a set of pairs, X/t in which X is a variable and t is a wat, and X occurs only once in the set. The *domain* of a substitution σ, written $Dom(\sigma)$ is the set $\{X \mid X/t \in \sigma\}$. Application of a well-annotated substitution σ to a well-annotated term t, written $\sigma(t)$, is as normal except term replacement is performed with our new subterm replacement function. Finally, if σ is a *was*, we define $erase(\sigma)$, and $skel(\sigma)$ as the result of applying these functions to each element in the range of the substitution, i.e.,

$$erase(\sigma) = \{X/t' \mid X/t \in \sigma \wedge t' = erase(t)\} .$$

The following can be easily verified by induction over the structure of *wats*.

LEMMA 1. *For σ a well-annotated substitution we have*

$$\sigma(\boxed{f(\underline{t_1}, \ldots, t_n)}^{\uparrow}) = \boxed{f(\underline{\sigma(t_1)}, \ldots, erase(\sigma)(t_n))}^{\uparrow}$$
$$= \boxed{f(\underline{\sigma(t_1)}, \ldots, erase(\sigma(t_n)))}^{\uparrow}$$

and

$$\sigma(\boxed{f(\underline{t_1}, \ldots, t_n)}^{\downarrow}) = \boxed{f(\underline{\sigma(t_1)}, \ldots, erase(\sigma)(t_n))}^{\downarrow}$$
$$= \boxed{f(\underline{\sigma(t_1)}, \ldots, erase(\sigma(t_n)))}^{\downarrow}.$$

Annotated substitution preserves well-formedness, skeletons, and erasure in the following sense:

THEOREM 2. *If t is a* wat *and σ is a well-annotated substitution, then*
1. $\sigma(t)$ is a wat,
2. $skel(\sigma(t)) = (skel(\sigma))(skel(t))$,
3. $erase(\sigma(t)) = (erase(\sigma))(erase(t))$.

Proof. (*sketch*) The proof follows by structural induction on t. As before, the only interesting case occurs when $t = \boxed{f(\underline{t_1}, \ldots, \underline{t_j}, t_{j+1}, \ldots, t_n)}^{\uparrow}$ (the inwards case occurs analogous). (1) follows as

$$\sigma(t) = \boxed{f(\underline{\sigma(t_1)}, \ldots, \underline{\sigma(t_j)}, erase(\sigma(t_{j+1})), \ldots, erase(\sigma(t_n)))}^{\uparrow}$$

and subterms in wave-fronts; that is, $\sigma(t_i)$ for $i \in \{1, \ldots, j\}$ are *wats* by the induction hypothesis and the remaining subterms are unannotated and therefore also *wats*. (2) follows as

$$\begin{aligned}
&skel(\sigma(t)) \\
&\quad = skel(\boxed{f(\underline{\sigma(t_1)}, \ldots, \underline{\sigma(t_j)}, erase(\sigma(t_{j+1})), \ldots, erase(\sigma(t_n)))}^{\uparrow}) \\
&\quad = \bigcup_{i=1}^{j} \{s \mid s \in skel(\sigma(t_i))\} \\
&\quad = \bigcup_{i=1}^{j} \{s \mid s \in skel(\sigma)(skel(t_1))\} \\
&\quad = (skel(\sigma))skel(\boxed{f(\underline{t_1}, \ldots, \underline{t_j}, t_{j+1}, \ldots, t_n)}^{\uparrow})
\end{aligned}$$

Finally, (3) follows as

$$erase(\sigma(t)) = erase(\boxed{f(\underline{\sigma(t_1)}, \ldots, erase(\sigma(t_n)))}^{\uparrow})$$

$$
\begin{aligned}
&= f(erase(\sigma(t_1)), \ldots, erase(erase(\sigma(t_n)))) \\
&= f(erase(\sigma(t_1)), \ldots, erase(\sigma(t_n))) \\
&= f((erase(\sigma))(erase(t_1)), \ldots, (erase(\sigma))(erase(t_n))) \\
&= (erase(\sigma))(f(erase(t_1), \ldots, erase(t_n))) \\
&= (erase(\sigma))(erase(t)). \qquad \Box
\end{aligned}
$$

To perform rewriting, we need a notion of annotated matching corresponding to this new notion of annotated substitution.

DEFINITION 5. If s and t are *wats*, then σ is an *annotated match* of s with t iff $Dom(\sigma) = Vars(s)$, σ is a *was*, and $\sigma(s) = t$.

Observe that even restricting σ to variables in s, annotated matching is not unitary. For example, $\{X/s(0)\}$ and $\{X/\boxed{s(\underline{0})}^{\uparrow}\}$ are both annotated matches of $\boxed{X \times \underline{0}}^{\uparrow}$ with $\boxed{s(0) \times \underline{0}}^{\uparrow}$. It can be seen, however, that matches differ only in the amount of annotation that appears on substitutions for variables that occur in wave-fronts but not in skeletons. Based on this observation, we can define a notion of minimality of annotation so that annotated matching is unique. If σ and τ are well-annotated substitutions for s and t, then we write $\sigma \prec \tau$ iff there exists $X/t_1 \in \sigma$ and $X/t_2 \in \tau$ with $t_1 = erase(t_2)$ and all other pairs in σ and τ are identical. We write $\prec^+$ for the transitive (but not reflexive) closure of $\prec$.

DEFINITION 6. If s and t are wats, then σ is a *minimal match* of s with t iff σ is an annotated match of s with t and there does not exist any annotated match τ with $\tau \prec^+ \sigma$.

It follows from this definition that if we have a minimal match, we cannot remove any annotation and have the result remain a match. We now give an algorithm for computing minimal matches. We will use this afterwards to show that minimal matches are unique.

Our procedure, $amatch(s, t)$, is based on the transformation rules given in Figure 1. Because term replacement, and hence substitution, is dependent on context (i.e., whether or not the term to be replaced is in a wave-front), our rules manipulate equations labeled with context information (wf for 'in the wave-front' and sk for 'in the skeleton'). As notational shorthand, *Pos* is a meta-variable that matches either wf or sk.

Starting with the set containing the match problem $\{s = t\!: sk\}$, we apply these transformation rules exhaustively. An equation set is *normalized* if no more transformation rules can be applied. A normalized equation set, S, is *compatible* iff (1) every equation is a variable assignment $X = s\!: Pos$ where s is a *wat*, (2) for each variable X there exists at most one equation of the form $X = s\!: sk$, and (3) if $X = s\!: sk \in S$ and $X = t\!: wf \in S$, then $erase(s) = t$. If a normalized

DELETE:
$$S \cup \{t = t\!:\!Pos\} \Rightarrow S$$

DECOMPOSE:
$$S \cup \{f(s_1, \ldots, s_n) = f(t_1, \ldots, t_n)\!:\!Pos\}$$
$$\Rightarrow S \cup \{s_i = t_i\!:\!Pos \mid 1 \leqslant i \leqslant n\}$$
$$S \cup \{\boxed{f(\underline{s_1}, \ldots, \underline{s_j}, s_{j+1}, \ldots, s_n)}^{\uparrow} = \boxed{f(\underline{t_1}, \ldots, \underline{t_j}, t_{j+1}, \ldots, t_n)}^{\uparrow}\!:\!sk\} \Rightarrow$$
$$S \cup \{s_1 = t_1\!:\!sk, \ldots, s_j = t_j\!:\!sk, s_{j+1} = t_{j+1}\!:\!wf, \ldots, s_n = t_n\!:\!wf\}$$
$$S \cup \{\boxed{f(\underline{s_1}, \ldots, \underline{s_j}, s_{j+1}, \ldots, s_n)}^{\downarrow} = \boxed{f(\underline{t_1}, \ldots, \underline{t_j}, t_{j+1}, \ldots, t_n)}^{\downarrow}\!:\!sk\} \Rightarrow$$
$$S \cup \{s_1 = t_1\!:\!sk, \ldots, s_j = t_j\!:\!sk, s_{j+1} = t_{j+1}\!:\!wf, \ldots, s_n = t_n\!:\!wf\}$$

Figure 1. Transformation rules for $amatch(s, t)$.

equation set is not compatible, matching has failed; otherwise matching has succeeded, and we return the answer substitution

$$\{X/s \mid X = s\!:\!sk \in S \text{ or } (X = s\!:\!wf \in S \text{ and } X = t\!:\!sk \notin S)\}.$$

As an example, if we match $\boxed{\underline{X} + X}^{\uparrow}$ with $\boxed{\underline{\boxed{s(\underline{a})}^{\uparrow}} + s(a))}^{\uparrow}$, then our initial matching problem is

$$\{\boxed{\underline{X} + X)}^{\uparrow} = \boxed{\underline{\boxed{s(\underline{a})}^{\uparrow}} + s(a))}^{\uparrow} : sk\}$$

and applying DECOMPOSE yields $\{X = \boxed{s(\underline{a})}^{\uparrow} : sk, X = s(a) : wf\}$. This normalized equation set is compatible and yields the answer $\{X/\boxed{s(\underline{a})}^{\uparrow}\}$. Note that regular matching would fail on this example.

It is easy to see that the application of these rules in any order terminates in time linear in the size of the smaller of s and t. Moreover, although there are choice points concerning which rule is applied, these choices are *AND* choices: all equations must be reduced, and each reduction is independent. It follows that when there is an answer substitution, it is unique. Below, when we show correctness, we demonstrate that the algorithm returns all minimal answers, so it follows that there is a unique such answer.

Note that the DELETE and the first DECOMPOSE rule implement regular matching. For unannotated terms, the compatibility check reduces to the requirement that the normalized equation set only contains variable assignments, and each variable has a single substitution. Annotated matching therefore subsumes regular matching. As with regular matching, we can also add two failure rules for greater efficiency: CONFLICT, which causes annotated matching to fail when the outermost function symbols disagree; and INCOMPATIBLE, which causes annotated matching to fail immediately if the set of equations of the form $X = s\!:\!Pos$ is not compatible. These additional failure rules are not, however, needed for the completeness or correctness of annotated matching.

THEOREM 3. *If s and t are* wats, *then*
1. *$amatch(s,t) = \sigma$ iff σ is the minimal match of s and t;*
2. *$amatch(s,t)$ fails iff s and t do not have an annotated match.*

Proof. (*sketch*) We prove a stronger result: A set of labeled equations S can be transformed to a compatible set of equations from which we extract σ iff for all $s = t\!:\!sk \in S$, $\sigma(s) = t$, and for all $s = t\!:\!wf \in S$, $(erase(\sigma))(s) = t$. The minimality of the answer substitution extracted holds by construction.

($\Rightarrow$) We use induction on the length of the transformation. If S is already a normalized compatible set of equations, then the result follows directly from the way the answer substitution is computed. Alternatively, we must apply a transformation rule. The interesting case is when DECOMPOSE is applied, say to $\boxed{f(\underline{s_1}, \ldots, s_n)}^{\uparrow} = \boxed{f(\underline{t_1}, \ldots, t_n)}^{\uparrow} : sk$, giving the set of equations S'. By the induction hypothesis, for all $s = t\!:\!sk \in S'$, $\sigma(s) = \sigma(t)$, and for all $s = t\!:\!wf \in S'$, $(erase(\sigma))(s) = (erase(\sigma))(t)$. By the definition of annotated substitution, we have $\sigma(\boxed{f(\underline{s_1}, \ldots, s_n)}^{\uparrow}) = \boxed{f(\underline{t_1}, \ldots, t_n)}^{\uparrow}$.

($\Leftarrow$) We use induction on the height of the largest LHS of an equation. If this is atomic, then, possibly after applications of DELETE, the equation set will be compatible. If it is not atomic, then the equation with the largest LHS will be of the form $f(s_1, \ldots, s_n) = t\!:\!Pos$, $\boxed{f(\underline{s_1}, \ldots, s_n)}^{\uparrow} = t\!:\!sk$, or $\boxed{f(\underline{s_1}, \ldots, s_n)}^{\downarrow} = t\!:\!sk$. In all three cases, we can apply DECOMPOSE and appeal to the induction hypothesis.

Note that the normalized equation set is not compatible iff either two occurrences of a variable in the skeleton need a different substitution, or a variable in the wave-front needs a substitution which is not the erasure of the substitution needed by an occurrence in the skeleton, or there is a conflict in function symbols or annotation preventing application of DECOMPOSE. But this occurs iff s and t do have an annotated match. □

4.3. First-Order Annotated Rewriting

We now consider rippling with proper rewrite rules that may contain variables. First, we define non-ground rippling. Let $s[t]$ be a *wat* with a distinguished subterm t and $l \rightarrow r$ be a proper rewrite rule. Further, let $\sigma = amatch(l, t)$.

Then $s[t]$ rewrites to $s[\sigma(r)]$, which we write as $s[\sigma(l)] \mapsto s[\sigma(r)]$. Correctness parallels the ground case. The proof relies on the fact that σ is a well-annotated substitution, and we can thus reduce the problem to the correctness of ground rippling.

THEOREM 4. *If s is a* wat, *$l \to r$ a proper rewrite rule, and $s[\sigma(l)] \mapsto s[\sigma(r)]$, then*

1. *$s[\sigma(r)]$ is a* wat,
2. $skel(s[\sigma(r)]) \subseteq skel(s[\sigma(l)])$,
3. $erase(s[\sigma(l)]) \to erase(s[\sigma(r)])$.

Proof. Annotated matching guarantees that σ is a *was*. By Theorem 2, $\sigma(l)$ and $\sigma(r)$ are *wats*, $skel(\sigma(r)) \subseteq skel(\sigma(l))$ and $erase(\sigma(l)) \to erase(\sigma(r))$. Now, since $\sigma(l)$ is syntactically identical to a subterm of s, it is ground (no non-term variables occur in s). Furthermore $\sigma(r)$ is also ground because $l \to r$ is a rewrite rule and thus $Vars(r) \subseteq Vars(l)$. Hence the rewriting of $s[\sigma(l)] \mapsto s[\sigma(r)]$ is equivalent to rewriting $s[\sigma(l)]$ with the ground proper rewrite rule $\sigma(l) \to \sigma(r)$. Thus, by Theorem 1, the three properties hold. □

Let $\mapsto^*$ be the reflexive transitive closure of $\mapsto$. By induction on the number of rewrite steps, it follows from Theorem 4 that rippling is correct and preserves skeletons. That is, if we erase annotations, we can perform the same (object-level) rewriting; annotations merely guide rewriting in a skeleton preserving way.

THEOREM 5. *If s is a* wat *and $s \mapsto^* t$, then*

1. *t is a* wat,
2. $skel(t) \subseteq skel(s)$,
3. $erase(s) \to^* erase(t)$.

4.4. TERMINATION

We now have a calculus that satisfies the first three properties required of rippling. Now we address the fourth: termination. As is conventional, we guarantee termination by placing an additional restriction on (proper) rewrite rules: they must reduce the measure of the rewritten term in a well-founded ordering.

We first define a weak kind of monotonicity that ensures monotonicity with respect to replacement of *wats* by *wats*. This is adequate to show termination when our rewrite rules are between *wats*.

DEFINITION 7. An order $>$ is *monotonic with respect to* wats iff for all *wats* s and for all ground *wats* l and r, if $l > r$, then $s[l] > s[r]$.

We now consider termination restricted to the ground case.

LEMMA 2. *Let R be a set of (not necessarily proper) rewrite rules between ground* wats. *If $>$ is well founded and monotonic with respect to* wats, *and for all $l \to r \in R$ we have $l > r$, then rippling with R is terminating.*

Proof. By monotonicity, an infinite rewrite sequence $t_1 \mapsto t_2 \mapsto \cdots$ gives an infinite sequence of *wats* $t_1 > t_2 > \cdots$, contradicting the well-foundedness of $>$. □

For non-ground rippling we consider a restricted form of stability.

DEFINITION 8. An order $>$ is *stable* with respect to *wats* iff for all well-annotated substitutions σ and *wats* s and t where $s > t$, we have that $\sigma(s) > \sigma(t)$. An order on annotated terms that is well founded, monotonic, and stable with respect to *wats* is an *annotation reduction order*.

Note that annotation reduction orders are strictly weaker than normal reduction orders, which are monotonic and stable over all terms in the signature, as opposed to just well-annotated ones. We will call our orders simply *reduction orders* when no confusion can arise.

Based on these definitions, we can at last formally define a *wave-rule*.

DEFINITION 9. Let $>$ be an annotation reduction order. Then a proper rewrite rule $l \to r$ is a *wave-rule with respect to* $>$ iff $l > r$.

THEOREM 6. *For an annotation reduction order $>$ and R a set of wave-rules with respect to $>$, rippling using wave-rules in R is terminating.*

Proof. We again reduce the problem to the ground case. If $s \mapsto t$ using $l \to r$, then this is equivalent to rippling with a rewrite rule $\sigma(l) \to \sigma(r)$ between ground *wats*. Since $>$ is stable with respect to *wats*, $\sigma(l) > \sigma(r)$. By the termination of ground rippling, we have termination in the general case. □

Note that our proof is similar to the one given in [11] (corollary to Theorem 5) as we need not show that the ordering is a reduction ordering, but rather only monotonic and stable with respect to possible instances of the rewrite rules.

5. Annotation Orders

To prove the termination of rippling using Theorem 6, we need to define a suitable order on annotated terms. We begin with simply annotated terms, those whose wave-fronts have a single wave-hole. We then generalize to orders for terms with multihole annotation. The orders we define are similar, though simpler than, that given by Bundy et al. in [6]. We can order all the wave-rules given in [6] and admit wave-rules not possible in their setting (see Section 8).

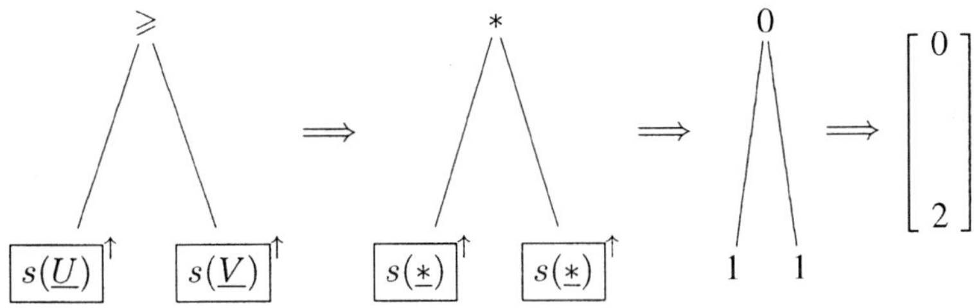

Figure 2. Defining a measure on annotated terms.

5.1. Single Wave-Holes

We consider annotated terms as decorated trees where the tree is the skeleton and the wave-fronts are boxes decorating the nodes. See, for example, the first tree in Figure 2, which represents the term $\boxed{s(\underline{U})}^{\uparrow} \geqslant \boxed{s(\underline{V})}^{\uparrow}$. Our orders are based on assigning measures to annotation in these trees. We define orders by progressively simplifying these annotated trees to capture the notion of progress during rippling that we wish to measure.

To begin with, since rippling is skeleton preserving, we need not account for the contents of the skeleton in our orderings. That is, we can abstract away function symbols in the skeleton, for example, mapping each to a variadic function constant '*'. This gives, for example, the second tree in Figure 2.

A further abstraction is to ignore the names of function symbols within wave-fronts and assign a numeric weight to wave-fronts. For example, we may tally up the values associated with each function symbol as in a Knuth–Bendix ordering. Two of the simplest kinds of weights that we may assign to wave-fronts measure their *width* and their *size*. Width is the number of nested function symbols between the root of the wave-front and the wave-hole. Size is the number of function symbols and constants in a wave-front. In what follows we will restrict our attention to the width measure. This gives, for example, the third tree in Figure 2. Of course, there are problem domains where we want our measure to reflect more of the structure of wave-fronts. Subsection 7.1 contains an example of this where the actual contents of the wave-front are compared using a conventional term ordering.

Finally, a very simple notion of progress during rippling is that wave-fronts move up or down through the skeleton tree. Under this view, the tree structure may be ignored: it is not important which branch a wave-front is on, only its depth in the skeleton tree. Hence, we can apply an abstraction that maps the tree onto a list, level by level. For instance, we can use the sum of the weights at a given depth. Applying this abstraction gives the final list in Figure 2. Note that depths are relative to the skeleton as opposed to depth in the erased term; measuring depth relative to a fixed skeleton is one of the key ideas in the measures proposed here.

To formalize the above ideas, we introduce the following definitions. As is standard, a *position* is simply a path address represented by a string and is defined as follows: the set of positions in the term t is $Pos(t)$, where

$$Pos(f(s_1, \ldots, s_n)) = \{\Lambda\} \cup \{i.p \mid 1 \leqslant i \leqslant n \ \wedge \ p \in Pos(s_i)\}$$

Λ represents the empty string and '.' is the string concatenation operator. The subterm of a term t at position p is t/p, where

$$\begin{aligned} t/\Lambda &= t \\ f(s_1, \ldots, s_n)/i.p &= s_i/p. \end{aligned}$$

If s is a subterm of t at position p, its *depth* is the length of p. The *height* of t, written $|t|$, is the maximal depth of any subterm in t.

Because we are interested in measures based on weight relative to the skeleton, during the remainder of this paper, positions, depth, and height will *always* be relative to the skeleton of simply annotated terms. That is, we picture such terms as in the first tree in Figure 2. The positions in the term tree are only those in the skeleton; annotation and function symbols in wave-fronts are treated as markings of function symbols in the skeleton. For example, the term in Figure 2 is $\boxed{s(\underline{U})}^{\uparrow} \geqslant \boxed{s(\underline{V})}^{\uparrow}$, which has skeleton $U \geqslant V$. The height of this term is 1, since the deepest subterms, U and V, have positions 1 and 2, respectively. Another example is $\boxed{f(s(\underline{f(a, s(b))}), c)}^{\uparrow}$ with the skeleton $f(a, s(b))$. The deepest subterm is b at position 2.1, and hence the height of the annotated term is 2.

For an annotated term t, the *out-weight of a position* p is the sum of the weights of the (possibly nested) outwards oriented wave-fronts at p. The *in-weight* is defined analogously except for inward directed wave-fronts. We now define a measure on terms corresponding to the final list in Figure 2 based on weights of annotation relative to their depths.

DEFINITION 10. The *out-measure*, $\mathcal{MO}(t)$ of an annotated term t is a list of length $|t| + 1$ whose ith element is the sum of out-weights for all term positions in t at depth i. The *in-measure*, $\mathcal{MI}(t)$ is a list whose ith element is the sum of in-weights for all term positions in t at depth i. The *measure* of an annotated term, $\mathcal{M}(t)$ is the pair of out and in-measures, $\langle \mathcal{MO}(t), \mathcal{MI}(t) \rangle$.

Consider, for example, the following palindrome function ('::' is infix cons)

$$palin(\boxed{H :: \underline{T}}^{\uparrow}, Acc) \rightarrow \boxed{H :: \underline{palin(T, \boxed{H :: \underline{Acc}}^{\downarrow})}}^{\uparrow} \tag{12}$$

The skeleton of both sides is $palin(T, Acc)$. The out-measure of the left-hand side is [0,1] and that of the right-hand side is [1,0]. The in-measures are [0,0] and [0,1].

We now define a well-founded ordering on these measures that reflects the progress that we want rippling to make. Consider, a simple wave-rule like (2),

$$\boxed{s(\underline{U})}^{\uparrow} \times V \rightarrow \boxed{(\underline{U \times V}) + V}^{\uparrow}.$$

The LHS out-measure is $[0, 1]$, and the RHS is $[1, 0]$. Rippling with this rule makes progress because it moves one wave-front upwards towards the root of the term. In general, rippling progresses if one out-oriented wave-front moves up or disappears, while nothing deeper moves downwards. If the out-measure of a term before rippling is $[l_0, \ldots, l_k]$ and after $[r_0, \ldots, r_k]$, then there must be some depth d where $l_d > r_d$ and for all $i > d$ we have $l_i = r_i$. This is simply the lexicographic order on the reverse of the two lists where components are compared using $>$ on the natural numbers. Progress for in-oriented wave-fronts is similar and reflects that these wave-fronts should move towards leaves, that is, just the lexicographic order on the in-measures. Of course, both outward and inward-oriented wave-fronts may occur in the same rule, e.g., (12). Similar to [6], we define a composite ordering on the out- and in-measures. We order the out-measure before the in-measure, since this procedure enables us to ripple wave-fronts out and either to reach the root of the term or at some point to turn wave-fronts down and to ripple in towards the leaves.

DEFINITION 11. $t \succ s$ iff $\mathcal{M}(t) > \mathcal{M}(s)$ and $skel(s) = skel(t)$. Here $>$ represents the lexicographic order on pairs with $>_{revlex}$ (the reversed lexicographic order on lists) used to compare the out-measure and $>_{lex}$ (the lexicographic order on lists) to compare the in-measure.

This definition is sensible because the restriction that $skel(s) = skel(t)$ means that the measure lists are the same length and may be compared. Although a skeleton – independent measure would be desirable – there is a deeper reason for this restriction: our order would not be stable without it. As a simple example, consider the terms $s = \boxed{X + \underline{s(s(Y))}}^{\uparrow}$ and $t = \boxed{s(\underline{X}) + Y}^{\uparrow}$. If we ignore the skeleton restriction and just compare annotation measures, then $s \succ t$. However, under the substitution $\sigma = \{X / \boxed{s(s(\underline{a}))}^{\uparrow}\}$ we have $\sigma(t) \succ \sigma(s)$. We will shortly show the stability of our more restricted ordering.

Given the well-foundedness of $>$ on the natural numbers and that lexicographic combinations of well-founded orders are well founded, we can conclude the following

THEOREM 7. *The composite ordering is well founded.*

5.2. Multiple Wave-Holes

We now generalize our order to multihole annotation; that is, multiple wave-holes in a single wave-front. Wave-rules involving such terms are called *multi-wave-rules* in [6]. We have already seen an example of this in (6). The binomial equation is another example.

$$binom(\boxed{s(\underline{X})}^{\uparrow}, \boxed{s(\underline{Y})}^{\uparrow}) = \boxed{\underline{binom(X, \boxed{s(\underline{Y})}^{\uparrow})} + \underline{binom(X,Y)}}^{\uparrow} \qquad (13)$$

Both sides have the same skeleton, namely, $\{binom(X,Y)\}$. In general, however, the skeletons of the right-hand side of a multi-wave-rule need only be a subset of the skeletons of the left-hand side.

We define orders for terms with multihole annotation in a uniform way from the previous single hole ordering by reducing terms with multihole annotation to sets of simply annotated terms and extending the single hole ordering to these sets. This reduction is accomplished by considering ways that multihole annotation can be *weakened* to simple annotation by 'absorbing' wave-holes. Weakening a multi-wave term like (13) erases some of the wave-holes (underlining), though always leaving at least one wave-hole. By erasing a wave-hole $\underline{t_i}$ we mean removing the underline annotation and erasing any further annotation in t_i. A wave-front is *maximally weak* when it has exactly one wave-hole. A term is *maximally weak* when all its wave-fronts are maximally weak. Maximally weak terms are simply annotated, and we can apply the previously defined single hole measure to them.

Returning to the binomial example, we see that (13) has precisely two weakenings.

$$binom(\boxed{s(\underline{X})}^{\uparrow}, \boxed{s(\underline{Y})}^{\uparrow}) = \boxed{\underline{binom(X, \boxed{s(\underline{Y})}^{\uparrow})} + binom(X,Y)}^{\uparrow} \qquad (14)$$

$$binom(\boxed{s(\underline{X})}^{\uparrow}, \boxed{s(\underline{Y})}^{\uparrow}) = \boxed{binom(X, s(Y)) + \underline{binom(X,Y)}}^{\uparrow} \qquad (15)$$

Both are maximally weak, since each wave-front has a single hole. As another example, the left-hand side of (6) has four maximal weakenings (and four non-maximal weakenings), while the right-hand side has two weakenings, both maximal.

Let $weakenings(s)$ be the set of maximal weakenings of s. It is easily computed by constructing the closure of all weakenings of s and returning the set of simply annotated results. As elements of these sets are simply annotated, we can apply the single hole measure to them. A natural order to define on such sets is therefore the multiset extension of the order used to compare simply annotated terms. A multiset extension of an ordering is defined as follows [11].

DEFINITION 12. A *multiset ordering* $>>$ is induced from a given ordering $>$ whereby $M >> N$ iff N can be obtained from M by replacing one or more elements in M by any finite number of elements each of which is smaller (under $>$) than one of the replaced elements.

We extend the single hole ordering to multihole annotated terms as follows.

DEFINITION 13. $l \succ^\star r$ iff $weakenings(l) \succ\!\succ weakenings(r)$ where $\succ\!\succ$ is the multiset extension of the single hole order $\succ$.

This order is well defined since maximal weakenings are simply annotated and can be compared by using the single-hole order. Note that if l and r are simply annotated, then their weakenings are $\{l\}$ and $\{r\}$, and $l \succ^\star r$ and $l \succ r$ are equivalent. We will drop the superscript on $\succ^\star$ when context makes our intention clear.

As an example consider (13). The LHS weakenings are

$$\{binom(\boxed{s(\underline{X})}^{\uparrow}, \boxed{s(\underline{Y})}^{\uparrow})\}\,.$$

The RHS weakenings are

$$\{\boxed{\underline{binom(X, \boxed{s(\underline{Y})}^{\uparrow})} + binom(X,Y)}^{\uparrow},$$

$$\boxed{binom(X + s(Y)) + \underline{binom(X,Y)}}^{\uparrow}\}\,.$$

The only member of the first set is $\succ$ greater than both members of the second set. This wave-rule is thus measure decreasing.

5.3. Termination under $\succ^\star$

Since $\succ^\star$ is defined via a multiset extension of a well-founded order, it too is well founded.

LEMMA 3. *$\succ^\star$ is well founded.*

We now show that $\succ^\star$ is monotonic and stable. To simplify proofs, we ignore complications caused by inwards oriented wave-fronts. Reincorporating these is conceptually simple but notationally involved, since measures expand to pairs.

As measures are lists, term replacement corresponds to operations on lists. Hence we begin with relevant terminology. Let l and r be list of integers and $l + r$ and $l - r$ be componentwise addition and subtraction. When one list is shorter than the other, we 'pad' it out by appending additional 0s to the end so that its length is the same as the longer. For n a natural number, let $l \uparrow^n$ be the result of 'right shifting' l by n positions by appending l to the end of the list containing n zeros. If the length of l is $n + 1$, then for any $d \in \{0, \ldots, n\}$ we define the *splice* of r into l at depth d, which we write as $l \; +_d \; r$, to be

$l + (r\uparrow^d)$. Splicing can result in a longer list; for example, if $l = [l_0, l_1, l_2, l_3]$ and $r = [r_0, r_1, r_2]$, then

$$\begin{aligned} l +_2 r &= l + (r\uparrow^2) = [l_0, l_1, l_2, l_3] + [0, 0, r_0, r_1, r_2] \\ &= [l_0, l_1, l_2 + r_0, l_3 + r_1, r_2]. \end{aligned}$$

We will use some simple properties about splice and list arithmetic below.

LEMMA 4. *Let l and l' be lists of length $i+1$ and $l >_{revlex} l'$. Let $r, r_1, \ldots, r_k$ be lists of length $j+1$. Then*

1. $\forall d \in \{0, \ldots, j\}. r >_{revlex} r +_d l' - l$
2. $\forall d \in \{0, \ldots, i\}. l +_d r >_{revlex} l' +_d r$
3. $\forall d_1, \ldots, d_m \in \{0, \ldots, i\}. (\cdots((l +_{d_1} r_1) +_{d_2} r_2) \cdots +_{d_m} r_l) >_{revlex} (\cdots((l' +_{d_1} r_1) +_{d_2} r_2) \cdots +_{d_m} r_l)$

The first lemma says we can splice in the difference between l' and l where $l >_{revlex} l'$ and the result will be smaller. The second says we can splice a list r into l and l' and preserve the ordering of l and l'. The third lemma is essentially an iterated version of the second for performing multiple splices with different lists at multiple positions. We use these results to prove theorems about stability and monotonicity, since such theorems can be seen as statements about splicing measures.

LEMMA 5. $\succ^\star$ *is monotonic with respect to* wats.

Proof. (*sketch*) Let $s[l]$ be a term with a distinguished subterm l. Note that if $l \succ r$, then l must be annotated. We argue by cases. Suppose first that s, l, and r are simply annotated. Let $m_l = \mathcal{MO}(l)$ be the measure of l; similarly let $m_r = \mathcal{MO}(r)$ and $m_s = \mathcal{MO}(s)$. Let d be the depth of l in the skeleton of s. The measure of $s[r]$ is the measure of s altered by splicing at depth d the difference between the measures of r and l, i.e., $m_s +_d (m_r - m_l)$. Since $l \succ r$, we can conclude, using the first part of Lemma 4, that $s[l] \succ^\star s[r]$.

Now suppose l and r contain multihole annotation and the only multihole annotation in $s[l]$ occurs in l itself. Let the maximal weakenings of l and r be the sets $L = \{l_1, \ldots, l_j\}$ and $R = \{r_1, \ldots, r_k\}$, respectively. The maximal weakenings of $s[l]$ and $s[r]$ then are the sets $S_l = \{s[l_1], \ldots, s[l_j]\}$ and $S_r = \{s[r_1], \ldots, s[r_k]\}$. Now under the definition of $\succ^\star$ and multisets, $l \succ^\star r$ if we can replace some collection of the $l_i \in L$ by smaller elements (under $\succ$) resulting in the set R. But we can do the identical replacements in the context $s[\cdot]$, hence transforming the set S_l to S_r. Consider such a replacement, say replacing $l_1 \in L$ by $r_1, \ldots, r_p$; now $l_1 \succ r_i$ and it follows (by the previously considered case) that $s[l_1] \succ s[r_i]$ for each $i \in \{1, \ldots, p\}$. Hence the transformation of S_l to S_r shows that $s[l] \succ^\star s[r]$.

The final case to consider is when s itself has multiple skeletons, independent of the number of skeletons of l. We argue as above except that rather than just comparing sets composed from $s[l_i]$ and $s[r_i]$ we have to consider weakings of s as well. But any steps in weakening s (not in the subterm l) can be made identically in both $s[l_i]$ and $s[r_i]$ and $s[l] \succ^{\star} s[r]$ follows. □

LEMMA 6. $\succ^{\star}$ *is stable with respect to* wats.

Proof. (*sketch*) Let s and t be *wats* with $s \succ^{\star} t$. To show that $\sigma(s) \succ^{\star} \sigma(t)$, it suffices to consider a substitution σ that replaces a single variable x with a wat r since substitution for multiple variables can be achieved by iterating substitutions for single variables. We consider two cases: first, when s, t are simply annotated, and second, when they may contain multihole wave-fronts.

Case 1: s and t are simply annotated. As $s \succ t$, both terms have the same single skeleton. Note that substitutions for occurrences of x in wave-fronts have no effect on our width measure (although they can change the size of a wave-front). Assume x occurs p times in each skeleton. If $weakenings(r) = \{r_1, \ldots, r_m\}$, then

$$S = weakenings(\sigma(s)) = \{s_1, \ldots, s_n\}$$

and

$$T = weakenings(\sigma(t)) = \{t_1, \ldots, t_n\}$$

where $n = p * m$. Each of these weakenings can be constructed by replacing the variables x in s and t with maximal weakenings of r; each s_i thus has a 'partner' t_i in which the occurrences of x are replaced by the same weakening of r. Now to show that S is greater than T under the multiset ordering, we must give a transformation of S to T where each term is replaced by a finite number of smaller (under $\succ$) terms. Our transformation is simply to replace s_i by its partner t_i. If we order (arbitrarily) the occurrences of x in the skeleton of s (and therefore also t), $x_1, \ldots, x_p$, then if s_i and t_i were formed by replacing x_j, occurring at depth d_j with a weakening of t that has a measure r_j, then the measures of the two terms s_i and t_i are

$$(\cdots((s +_{d_1} r_1) +_{d_2} r_2) \cdots +_{d_p} r_p)$$

and

$$(\cdots((t +_{d_1} r_1) +_{d_2} r_2) \cdots +_{d_p} r_p),$$

respectively. But now, using part 3 of Lemma 4, we have that the former is greater under $>_{revlex}$ than the latter, hence $\sigma(l) \succ \sigma(r)$.

Case 2: all terms may contain multihole annotation. Let $S = \{s_1, \ldots, s_j\}$ and $T = \{t_1, \ldots, t_k\}$ be the maximal weakenings of s and t. As $s \succ^{\star} t$, there is a transformation (respecting $\succ$) of S to T. We must construct a transformation from the maximal weakenings of $\sigma(s)$ to the maximal weakenings of $\sigma(t)$. We proceed as follows. Consider a replacement of, say, s_1 in S with some $t_1, \ldots, t_p$ that takes place in transforming S to T. Now suppose the maximal weakenings of r are $\{r_1, \ldots, r_m\}$. Then $\sigma(s_1)$ and the $\sigma(t_i)$ each has n maximal weakenings where n is a multiple of m dependent on the number of occurrences of x in the skeleton of s_1. In particular, $weakenings(\sigma(s_1)) = \{s_{1,1}, \ldots, s_{1,n}\}$ and for each t_i, $weakenings(\sigma(t_i)) = \{t_{i,1}, \ldots, t_{i,n}\}$. Again we may speak of 'partners': each $s_{1,j}$ has as partners $t_{i,j}$, for $i \in \{1, \ldots, p\}$ and $j \in \{1, \ldots, n\}$ where the weakenings of $t_{i,j}$ come from weakening the occurrences t identically to their weakenings in $s_{1,j}$. Furthermore, because for each $i \in \{1, \ldots, p\}$, $s_1 \succ t_i$, we can use Case 1 to conclude that each maximal weakening of $\sigma(s_1)$ is larger than its partners. Hence replacing each $s_{1,i}$ with its partners defines an appropriate transformation from $weakenings(\sigma(s))$ to $weakenings(\sigma(t))$. □

As $\succ^\star$ is an annotation reduction ordering we can conclude the following.

THEOREM 8. *Rippling using proper rewrite rules $l \to r$ for which $l \succ^\star r$ is terminating.*

6. Implementing Rippling

We have completed our development of a calculus for rippling and termination orderings for annotated terms. We now consider the more practical problem of mechanizing such a calculus. In particular, given an ordering, how do we then recognize wave-rules and apply them? We have implemented the rewrite calculus described, and here we indicate how the simplicity of our calculus and orderings led to a simple implementation, which now makes up part of the Edinburgh CLAM system. To give the reader a feel for this, and the issues involved, we briefly sketch a couple of the core routines.

Much of the work in implementing rippling concerns turning unannotated rewrite rules into wave-rules; we call this *wave-rule parsing*. A wave-rule parser must, given unannotated rewrite rules, return wave-rules, that is a collection of annotated copies of the rule that are proper rewrite rules and measure decreasing. We can achieve the requirements of proper annotation and measure decreasingness separately. An *annotation phase* first annotates l and r with unoriented wave-fronts so their skeletons are identical; this guarantees that rippling is skeleton preserving. An *orientation phase* then orients the wave-fronts with up and down arrows so that $l \succ r$. We sum this up by the slogan

$$\text{WAVE-RULE} = \text{ANNOTATION} + \text{ORIENTATION}.$$

As an example, consider parsing a rewrite rule like

$$s(U) \times V \to (U \times V) + V. \tag{16}$$

We may proceed by annotating this so the two sides have identical skeletons, i.e.,

$$\boxed{s(\underline{U})} \times V \to \boxed{(\underline{U \times V}) + V}. \tag{17}$$

Afterwards we can orient the annotation, yielding the wave-rule

$$\boxed{s(\underline{U})}^{\uparrow} \times V \rightarrow \boxed{(\underline{U \times V}) + V}^{\uparrow}. \tag{18}$$

Both sides of (12) now have the same skeleton, and the measure of the left-hand side is greater than that of the right-hand side.

Any implementation, however, must cope with the problem that under our definition of wave-rules, a given rewrite rule can generate exponentially many (in the size of the input rule) wave-rules. Computing and storing all possible wave-rules is expensive both in time and space and complicates efficient wave-rule lookup. For example, in the previous example, there are other possible legal parsing such as

$$\boxed{s(\underline{U}) \times V}^{\uparrow} \rightarrow \boxed{\underline{U} \times V + V}^{\downarrow} \tag{19}$$

$$\boxed{s(U) \times \underline{V}}^{\uparrow} \rightarrow \boxed{U \times \underline{V} + V}^{\downarrow} \tag{20}$$

$$\boxed{s(U) \times \underline{V}}^{\uparrow} \rightarrow \boxed{U \times V + \underline{V}}^{\downarrow}. \tag{21}$$

These additional parsings are problematic; while they are not really in the 'spirit' of wave-rules as originally proposed by [6] (nor admissible under their definition) and are seldomly useful in practice, they are admissible under our more liberal definition and on occasion find use in, for example, wave-front normalization (we discuss this in Subsection 7.1). Rather than trying to say in advance which wave-rules could be useful in practice, our solution to this problem is to compute wave-rules dynamically, by parsing 'on demand'. We describe this in the following section.

6.1. Dynamic Wave-Rule Parsing

We have implemented a *dynamic parser* that, given a data-base of unannotated rewrite rule, uses them for rippling by annotating them only as required during rewriting. That is, given a term $s[t]$ to be rewritten, we look for an unannotated rule $l \rightarrow r$ where l matches the erasure of t. When this is the case, $l \rightarrow r$ is a candidate wave-rule. We proceed by computing annotations for l that allow for an annotated match with t; afterwards, based on these annotations and their orientations, we compute annotations and orientations for r so that $l \rightarrow r$ is a wave-rule.

Figure 3 contains the actual Prolog (we assume the reader is familiar with Prolog in the following) that implements the top level 'architecture' of rippling based on dynamic parsing. We illustrate the procedure through an example. Suppose we wish to perform one step of rewriting of the term $T = \boxed{s(\underline{x})}^{\uparrow} \times \boxed{s(\underline{y})}^{\uparrow}$ using a collection of unannotated rewrite rules that includes (16), the recursive definition of multiplication. `Rewrite` picks a subterm ST of T and a rule

```
rewrite(T,NT) :-                   % rewrite at some term position
  subterm(Pos,T,ST),               % find a subterm ST in T
  pick_rule(L,R),                  % pick a rule L -> R
  match_rule(L,R,ST,NR),           % can rule be annotated to match ST
  replace(Pos,T,NR).               % replace subterm ST with NR

match_rule(L,R,ST,NR) :-
  copy_an(ST,L,AL),                % copy annotation from ST onto L
  amatch(AL,ST,Sigma),             % annotated match of AL with ST
  parse(AL,R,AR),                  % find annotations for R
  apply_subs(Sigma,AR,NR).         % apply substitution to AR

parse(AL,R,AR) :-
  pick_an(R,A),                    % annotate R
  skel_preserving(AL,A),           % skeletons equal?
  orient(AL,A,AR).                 % Orient R
```

Figure 3. Wave-rule parser (top-level routines).

$L \to R$. In our example, a solution to this (there may be others that are returned on backtracking) is where ST is T itself and the rule selected is (16). In dynamic parsing we need only generate annotations for the right-hand sides of wave-rules whose left-hand sides arise during the proof. This is performed in `match_rule`, which starts by copying annotations from ST onto L; this yields AL, an annotated version of L. Copying annotation fails if ST and L have erasures that do not match. In our example, $AL = \boxed{s(\underline{U})} \times V$. We then annotated match AL with ST to find a suitable substitution for the rewriting step. In our example, we get the substitution $\{U/x,\ V/\boxed{s(\underline{y})}^{\uparrow}\}$. After, we call `parse` to find an annotation of R with the same skeleton as AL, and with a maximum[2] orientation, in this case $\boxed{(\underline{U \times V}) + V}^{\uparrow}$. We then apply the substitution to this annotated right-hand side, giving $\boxed{(\underline{x \times \boxed{s(\underline{y})}^{\uparrow}}) + s(y)}^{\uparrow}$. Note that substitution application uses our new notion of subterm replacement (from Section 4). Here, regular substitution would generate an improperly annotated term. The final step in rewriting is to replace the rewritten subterm within the context of the superterm from which it came, again using our new notion of subterm replacement.

The supporting subroutines for parsing are fairly straightforward. We used generate (`pick_an`) and test (`skel_preserving`) to generate skeleton preserving annotations of the right-hand side of rewrite rules. This takes exponential time but the efficiency can be considerably improved by interleaving generation and testing (i.e., testing during generation) or via dynamic programming. A dynamic programming approach to such *difference matching* problems is described in [1]. In our experience, naive generate and test has acceptable performance.

The routine `orient` finds an orientation of the wave-fronts on the right-hand side that yields a measure smaller than the left-hand side. This can be implemented naively by generating orientations (there are two possibilities for each wave-front) and comparing the two sides of the proposed rule under the given measure. By comparing possible orientations against each other, we can return the maximum possible right-hand side orientations. As with annotation, there are algorithms to implement orientation more efficiently. In particular, when all annotation is simple (single wave-holes), it is possible to orient the right-hand side in linear time (in the size of the term). An algorithm for this is given in [2].

6.2. Sinks and Colors

One kind of annotation we have not discussed in our measures or parsing is *sinks*. This is deliberate, since we can safely ignore sinks in both the measure and the parser. Sinks serve only to decrease the applicability of wave-rules by creating additional preconditions; that is, we ripple inwards only if there is a sink underneath the wave-front. Hence, sinks decrease the search space of rippling, and termination without this restriction implies termination with this restriction. The value of sinks is they restrict search without reducing the *utility* of rippling: their use guides rippling in a way that allows the induction hypothesis to be successfully applied.

Another type of annotation introduced in [17] are wave-holes marked with colors. Different colors are used to distinguish different skeletons. Colors are useful in inductive proofs with multiple induction hypotheses (for example, inductions on binary trees). The motivation behind the introduction of colors is that rippling preserves only a subset of the skeletons, and colors help prevent us ending up with the wrong subset. Since colored rippling is a restriction of uncolored rippling, termination follows immediately from termination in the uncolored case. Colors thus increase the utility of rippling. Although colors are not needed for showing the termination of rippling, they actually played an implicit role in our discussion about termination. The reduction order defined in Section 5 compares the measures of different colors separately. Since each color has a single skeleton, and the measure is stable for single skeletons, the resulting order is stable.

7. Extensions to Rippling

Our definition of wave-rules is parameterized by a reduction ordering. This gives us flexibility in exploring new orderings and hence new applications for rippling. To illustrate this potential, we give two examples for inductive and non-inductive theorem proving. The first highlights a problem occurring in induction: during rippling we may need to normalize the contents of wave-fronts when rippling gets stuck. The second explores orderings useful for algebraic problem solving.

7.1. Unblocking

Here we consider new reduction orderings motivated by unblocking rippling. That is, sometimes rippling fails because no wave-rule is applicable, but not all wave-fronts have been moved 'out of the way' (to the root of the term or to sinks). This can occur because a lemma is needed; these missing wave-rules can sometimes be speculated automatically using techniques presented by Ireland and Bundy in [15]. Rippling can also become blocked simply because a wave-front itself needs to be rewritten so that it matches either an existing wave-rule (to allow further rippling) or a sink (to allow use of the induction hypothesis). This is best illustrated by an example taken from [6].

Consider the following theorem, where rev is naive reverse, $qrev$ is tail-recursive reverse using an accumulator, $<>$ is infix append, and $::$ infix cons.

$$\forall L, M.\ qrev(L, M) = rev(L) <> M \tag{22}$$

We proceed by induction on L. The induction hypothesis is

$$qrev(l, M) = rev(l) <> M,$$

where M is a universally quantified variable. The induction conclusion is

$$qrev(\boxed{h :: \underline{l}}^{\uparrow}, \lfloor m \rfloor) = rev(\boxed{h :: \underline{l}}^{\uparrow}) <> \lfloor m \rfloor, \tag{23}$$

where m is a Skolem constant that sits in a sink, annotated with '$\lfloor\ \rfloor$'.

We use wave-rules taken from the recursive definition of $qrev$, and rev.

$$rev(\boxed{H :: \underline{T}}^{\uparrow}) \rightarrow \boxed{\underline{rev(T)} <> (H :: nil)}^{\uparrow} \tag{24}$$

$$qrev(\boxed{H :: \underline{T}}^{\uparrow}, L) \rightarrow qrev(T, \boxed{H :: \underline{L}}^{\downarrow}) \tag{25}$$

On the LHS, we ripple with (25) to give

$$qrev(l, \left\lfloor \boxed{h :: \underline{m}}^{\downarrow} \right\rfloor) = rev(\boxed{h :: \underline{l}}^{\uparrow}) <> \lfloor m \rfloor .$$

The sink stays in the same position relative to the skeleton and absorbs the wave-front rippled across by (25). On the RHS, we ripple with (24) and then (8), the associativity of $<>$, to get

$$qrev(l, \left\lfloor \boxed{h :: \underline{m}}^{\downarrow} \right\rfloor) = rev(l) <> (\left\lfloor \boxed{(h :: nil) <> \underline{m}}^{\downarrow} \right\rfloor). \tag{26}$$

Again note how the sink has absorbed the wave-front rippled across. Unfortunately, the proof is now blocked. We cannot ripple any further nor apply the induction hypothesis. The problem is that we need to simplify the wave-front on the right-hand side so that the two sinks become identical. CLAM currently uses an ad hoc method to try to perform wave-front simplification when rippling becomes blocked. In this case (26) is rewritten by using conventional rewriting to

$$qrev(l, \left\lfloor \boxed{h :: \underline{m}}^{\downarrow} \right\rfloor) = rev(l) <> (\left\lfloor \boxed{h :: \underline{m}}^{\downarrow} \right\rfloor).$$

Simplification with the induction hypothesis can now occur.

Unblocking steps that manipulate just wave-fronts will use proper rewrite rules; for example, here we use another parsing for the recursive definition of append.

$$\boxed{(H :: T) <> \underline{L}}^{\downarrow} \rightarrow \boxed{H :: (T <> \underline{L})}^{\downarrow} \tag{27}$$

In [6] such a rule is not admitted as a wave-rule (see Section 8). It is also not admissible under our ordering $\succ^*$ as $\succ^*$ measures the *width* of wave-fronts and the right-hand side is wider than the left-hand side. However, both sides have the same size (number of function symbols and constants). If we extend our measure to account for the contents of wave-fronts, we can find a reduction ordering based on size of wave-fronts that includes the above rule.

We proceed as follows. As before, we give an ordering on simply annotated terms, which can then be lifted to an order on multiwave terms. To order simply annotated terms, we take the lexicographic order of the simple wave-rule measure proposed above (using size of the wave-front as the notion of weight) paired with $>_{wf}$, an order on the *contents* of wave-fronts. As a simply annotated term may contain multiple wave-fronts, this second order is lifted to a measure on sets of wave-fronts by taking its multiset extension. The first part of the lexicographic ordering ensures that anything that is normally measure decreasing remains measure decreasing, and the second part allows us to orient rules that only manipulate wave-fronts. This combination can provide a reduction ordering that allows us to use rippling to move wave-fronts about the skeleton and conventional rewriting to manipulate the contents of these wave-fronts.

In our reverse example, (27) does not change the size of the wave-front or its position but only its form. Hence we want this to be decreasing under some ordering on the contents of wave-fronts. There are many such orderings; here we take $>_{wf}$ to be the recursive path ordering [12] on the terms in the wave-front where $<>$ has a higher precedence than $::$ and all other function symbols have an equivalent but lower priority. The measure of the LHS of (27) is now greater than that of the RHS as its wave-front is $(H :: T) <> *$ which is greater than $H :: (T <> *)$ in the recursive path ordering (to convert wave-fronts into well-formed terms, wave-holes are marked with the new symbol $*$).

Unblocking steps which simplify wave-fronts are useful in many proofs enabling both immediate application of the induction hypothesis (as in this example) and continued rippling. By defining new orders we can combine rippling with conventional term rewriting so that rules to rewrite wave-fronts are measure decreasing wave-rules accepted by the parser and applied like other wave-rules.

7.2. Algebraic Problem Solving

Rippling has found several novel uses besides inductive theorem proving. For example, it has been used to sum series [16], to prove limit theorems [17], and to guide equational reasoning [10]. However, new domains, especially non-inductive ones, require new orderings to guide proof construction. Here we sketch an application based on the PRESS system [9].[3] To solve algebraic equations, PRESS uses a set of methods that apply rewrite rules. The three main methods are *isolation*, *collection*, and *attraction*. Below are examples of rewrite rules used by these methods.

$$\begin{array}{lrcl} \text{ATTRACTION}: & \boxed{\log(\underline{U}) + \log(\underline{V})}^{\uparrow} & \rightarrow & \boxed{\log(\underline{U} \times \underline{V})}^{\uparrow} \\ \text{COLLECTION}: & \boxed{\underline{U} \times \underline{U}}^{\uparrow} & \rightarrow & \boxed{\underline{U}^2}^{\uparrow} \\ \text{ISOLATION}: & \boxed{\underline{U}^2}^{\uparrow} = V & \rightarrow & U = \boxed{\pm\sqrt{\underline{V}}}^{\downarrow} \end{array}$$

PRESS uses preconditions and not annotation to determine rewrite rule applicability. Attraction must bring occurrences of unknowns closer together. Collection must reduce the number of occurrences of unknowns. Finally, isolation must make progress towards isolating unknowns on the LHS of the equation. These requirements can be captured by annotation, and PRESS can thus be implemented by rippling. The above wave-rules suggest how this would work. The wave-rules in PRESS are structure preserving, where the preserved structure is the unknowns. The ordering used reflects the well-founded progress achieved by the PRESS methods. Namely, we lexicographically combine orderings on the number of wave-holes for collection, their distance (shortest path between wave-holes in term tree) for attraction, and our width measure on annotation weight for isolation.

8. Related Work

8.1. CLAM

Our starting point is rippling as developed at Edinburgh and implemented in the CLAM proof planning system. Our results improve those presented in [6] in a number of respects.

To begin with, rippling as described in [6] is not a rewriting calculus. Rather it is implemented by first-order rewriting with the strong precondition that "... each wave-front in the expression [being rewritten] is matched with a wave-front of the same kind in the rule" (Definition 5, p. 222). Saying this another way, variables in wave-rules cannot be instantiated with annotated terms. This is sufficient for rippling to be structure preserving and terminating, but it is an unacceptably large restriction on the applicability of rippling. Indeed, under this restriction, not all of the examples in [6] are valid. For example (see p. 222) we cannot rewrite the immediate subterm of

$$even(\boxed{s(\underline{\boxed{s(\underline{x})}^{\uparrow}})}^{\uparrow} + y)$$

with the recursive definition of plus given by $\boxed{s(\underline{U})}^{\uparrow} + V \rightarrow \boxed{s(\underline{U + V})}^{\uparrow}$, since the left-hand side of this wave-rule is $\boxed{s(\underline{U})}^{\uparrow} + V$ and there is an extra wave-front in the subterm being rewritten.

Rippling was implemented in the CLAM system without the above restriction, and it suffered from the problems described in Section 3 that arise when first-order rewriting is used to implement rippling directly. In particular, ill-formed terms appeared during rewriting, and an auxiliary routine occasionally would 'clean-up' annotations (e.g., consider the multiplication example given in Section 3). The CLAM implementation of rippling has been replaced with the calculus and parser described here.

The measures and orders we give are considerably simpler than those in [6], where the properties of structure preservation and the reduction of a measure are intertwined. Bundy et al. describe wave-rules schematically and show that any instance of these schemata is skeleton preserving and measure decreasing under an appropriately defined measure. Mixing these two properties makes the definition of wave-rules very complex. For example, the simplest kind of wave-rule proposed are *longitudinal* outward directed wave-rules, defined as rules of the form

$$\eta(\boxed{\xi_1(\underline{\mu_1^1}, \ldots, \underline{\mu_1^{p_1}})}^{\uparrow}, \ldots, \boxed{\xi_n(\underline{\mu_n^1}, \ldots, \underline{\mu_n^{p_n}})}^{\uparrow})$$
$$\rightarrow \boxed{\zeta(\underline{\eta(\varpi_1^1, \ldots, \varpi_n^1)}, \ldots, \underline{\eta(\varpi_1^k, \ldots, \varpi_n^k)})}^{\uparrow},$$

which satisfy a number of side conditions. These include: each ϖ_i^j is either an unrippled wave-front, $\boxed{\xi_i(\underline{\mu_i^1}, \ldots, \underline{\mu_i^{p_i}})}$, or is one of the wave-holes, μ_i^l; for each j, at least one ϖ_i^j must be a wave-hole. η, the ξ_is, and ζ are terms with distinguished arguments; ζ may be empty, but the ξ_is and η must not be. There are other schemata for *transverse* wave-rules and *creational* wave-rules.[4] These schemata are combined in a general format, so complex that in [6] it takes four lines to print. It is notationally involved although not conceptually difficult to demonstrate that any instance of these schemata is a wave-rule under both our size and width measures.

Consider the longitudinal schema given above. Every skeleton on the LHS is a skeleton of the RHS because of the constraint on the ϖ^i_j. What is trickier to see is that it is measure decreasing. Under our order this is the case if LHS $\succ^*$ RHS. We can show something stronger, namely, for every $r \in \mathit{weakenings}(RHS).\, \exists l \in \mathit{weakenings}(LHS).\, l \succ r$. To see this, observe that any such r must be a maximal weakening of

$$r' = \boxed{\zeta(\eta(\varpi^1_1, \ldots, \varpi^1_n), \ldots, \underline{\eta(\varpi^j_1, \ldots, \varpi^j_n)}, \ldots \eta(\varpi^k_1, \ldots, \varpi^k_n))}^{\uparrow}$$

for some $j \in \{1, \ldots, k\}$. Corresponding to r' is an l' which is a weakening of the LHS where $l' = \eta(t_1, \ldots, t_n)$ and the t_i correspond to the ith subterm of $\eta(\varpi^j_1, \ldots, \varpi^j_n)$ in r': if ϖ^j_i is an unrippled wave-front then $t_i = \varpi^j_i = \boxed{\xi_i(\underline{\mu^1_i}, \ldots, \underline{\mu^{p_i}_i})}$, and alternatively if ϖ^j_i a wave-hole μ^l_i then $t_i = \boxed{\xi_i(\mu^1_i, \ldots, \underline{\mu^l_i}, \ldots, \mu^{p_i}_i)}$. Now r is a maximal weakening of r' so there is a series of weakening steps from r to r'. Each of these weakenings occurs in a ϖ^j_i, and we can perform the identical weakening steps in the corresponding t_i in l' leading to a maximal weakening l. l and r have the same skeleton, and since they are maximally weak, they may be compared under $\succ$. Their only differences are that r has an additional wave-front at its root, and is missing a wave-front at each ϖ^j_i corresponding to a wave-hole. The depth of ϖ^j_i is greater than the root, and at this depth the out-measure of l is greater than r, and at all greater depths they are identical. Hence $l \succ r$.

Similar arguments hold for the other schemata given in [6], and hence we can conclude that wave-rules acceptable under their definition are acceptable under ours. Moreover, simple examples are wave-rules under our formalism but not theirs, e.g., the base-case of addition $\boxed{0 + \underline{x}}^{\uparrow} \to x$.

8.2. INKA

Hutter, in [14, 13], describes a calculus for rippling implemented in the INKA system [3]. Hutter rigorously develops an algebra of annotated terms, called C-terms. These are terms in an extended signature where functions and variables each carry a 'color', which represents annotation, or a variable over colors, which restricts potential annotation. Hutter's motivations and developments are similar: he defines congruence relations corresponding to equality of terms after erasure, equivalence of skeletons, and develops algorithms to unify and rewrite C-terms that respect these congruences.

The calculus he develops is more general than ours. However, it is significantly more complex, both conceptually and in implementation. Wave-fronts can be thought of as contexts. In our calculus we augment the signature only as is required to specify these contexts: that is, we introduce new function symbols so that we may mark the beginning of a context with a wave-front, and the end of the context with wave-holes. In Hutter's calculus, annotation is the primary concept and matching, and rewriting of such terms can be understood independently of contexts.

Hutter has not addressed termination in his work. However, with minor restrictions on his calculus, our results should carry over. For example, we can consider a setting with three colors (indicating skeleton, inwards wave-fronts, and outwards wave-fronts) restricted to C-terms that are *wats*. In this setting we can define the same kinds of well-founded orderings on terms based on annotation relative to the skeleton. It should be possible to carry over our proofs of stability and monotonicity in his setting, although we have not formally checked this.

9. Conclusions

We have defined a simple calculus for rippling where differences between the induction conclusion and the induction hypothesis are marked with annotations and annotated rewrite rules move these differences away. We have proved that rewriting in this calculus has various desirable properties: well-formedness (well-annotated terms rewrite to well-annotated terms), skeleton preservation (the unannotated part of terms are preserved), and correctness (the corresponding rewriting can be performed in the original unannotated theory, i.e., annotation merely guides search). We have shown how this calculus admits simple termination orders which are stable and monotonic. As well as providing a firmer theoretical foundation for rippling, this work has led to a simpler and more complete implementation of rippling within the Edinburgh CLAM system.

Notes

[1] This generalization is, however, briefly discussed in their further work section.

[2] Maximum under our order. When there are multiple choices with the same measure we return them all on backtracking.

[3] The idea of reconstructing PRESS with rippling was suggested by Nick Free and Alan Bundy.

[4] Creational wave-rules are used to move wave-fronts between the hypotheses and conclusion during proofs by destructor induction. They complicate rippling in a rather specialized and uninteresting way. Our measures could be easily generalized to include these.

References

1. Basin, David and Walsh, Toby: Difference unification, in *Proc. 13th IJCAI*, International Joint Conference on Artificial Intelligence, 1993, pp. 116–122.
2. Basin, David and Walsh, Toby: Termination orderings for rippling, in *Proc. 12th Int. Conf. Automated Deduction (CADE-12)*, Nancy, France, June 1994, Springer-Verlag.

3. Biundo, Susanne, Hummel, Birgit, Hutter, Dieter, and Walther, Christoph: The Karlsruhe induction theorem proving system, in *8th Int. Conf. Automated Deduction*, Oxford, UK, 1986.
4. Boyer, Robert S. and Moore, J Strother: *A Computational Logic*. Academic Press, 1979.
5. Bundy, Alan, van Harmelen, Frank, Hesketh, Jane, and Smaill, Alan: Experiments with proof plans for induction, *J. Automated Reasoning* **7** (1991), 303–324.
6. Bundy, Alan, Stevens, Andrew, van Harmelen, Frank, Ireland, Andrew, and Smaill, Alan: Rippling: A heuristic for guiding inductive proofs, *Artificial Intelligence* **62** (1993), 185–253.
7. Bundy, Alan, van Harmelen, Frank, Horn, Christian, and Smaill, Alan: The Oyster-Clam system, in M. E. Stickel (ed.), *10th Int. Conf. Automated Deduction*, Lecture Notes in Artificial Intelligence 449, Springer-Verlag, 1990, pp. 647–648.
8. Bundy, Alan, van Harmelen, Frank, Smaill, Alan, and Ireland, Andrew: Extensions to the rippling-out tactic for guiding inductive proofs, in M. E. Stickel (ed.), *10th Int. Conf. Automated Deduction*, Lecture Notes in Artificial Intelligence 449, Springer-Verlag, 1990, pp. 132–146.
9. Bundy, Alan and Welham, Bob: Using meta-level inference for selective application of multiple rewrite rules in algebraic manipulation, *Artificial Intelligence* **16**(2) (1981), 189–212.
10. Cleve, Jürgen and Hutter, Dieter: A methodology for equational reasoning, in *HICSS-27*, IEEE, 1994.
11. Dershowitz, Nachum: Termination of rewriting, in J.-P. Jouannaud (ed.), *Rewriting Techniques and Applications*, Academic Press, 1987, pp. 69–116.
12. Dershowitz, Nachum: Orderings for term-rewriting systems, *Theor. Computer Science* **17**(3) (March 1982), 279–301.
13. Hutter, Dieter: Colouring terms to control equational reasoning, an expanded version of PhD Thesis: Mustergesteuerte Strategien für Beweisen von Gleichheiten, Universität Karlsruhe, 1991.
14. Hutter, Dieter: Guiding inductive proofs, in M. E. Stickel (ed.), *10th Int. Conf. Automated Deduction*, Lecture Notes in Artificial Intelligence 449, Springer-Verlag, 1990, pp. 147–161.
15. Ireland, Andrew and Bundy, Alan: Productive use of failure in inductive proof, *J. Automated Reasoning* **16**(1–2) (1996), 79–111.
16. Walsh, Toby, Nunes, Alex and Bundy, Alan: The use of proof plans to sum series, in D. Kapur (ed.), *11th Conf. Automated Deduction*, Lecture Notes in Computer Science 607, Springer-Verlag, 1992, pp. 325–339.
17. Yoshida, Tetsuja, Bundy, Alan, Green, Ian, Walsh, Toby, and Basin, David: Coloured rippling: An extension of a theorem proving heuristic, in *ECAI-94*, Wiley, New York, 1994.

Journal of Automated Reasoning **16:** 181–222, 1996.

Interaction with the Boyer–Moore Theorem Prover: A Tutorial Study Using the Arithmetic–Geometric Mean Theorem

MATT KAUFMANN*
Computational Logic, Inc., Austin, TX 78703, U.S.A. e-mail: kaufmann@cli.com
Current Address: Motorola, Inc., 5918 W. Courtyard Dr., Suite 330, Austin, TX 78730, U.S.A. e-mail: matt_Kaufmann@email.mot.com

PAOLO PECCHIARI
Mechanized Reasoning Group, IRST – Povo, 38100 Trento, Italy and DIST, University of Genoa, Genoa, Italy. e-mail: peck@itc.it

Abstract. There are many papers describing problems solved using the Boyer–Moore theorem prover, as well as papers describing new tools and functionalities added to it. Unfortunately, so far there has been no tutorial paper describing typical interactions that a user has with this system when trying to solve a nontrivial problem, including a discussion of issues that arise in these situations. In this paper we aim to fill this gap by illustrating how we have proved an interesting theorem with the Boyer–Moore theorem prover: a formalization of the assertion that the arithmetic mean of a sequence of natural numbers is greater than or equal to their geometric mean. We hope that this report will be of value not only for (non-expert) users of this system, who can learn some approaches (and tricks) to use when proving theorems with it, but also for implementors of automated deduction systems. Perhaps our main point is that, at least in the case of Nqthm, the user can interact with the system without knowing much about how it works inside. This perspective suggests the development of theorem provers that allow interaction that is user oriented and *not* system developer oriented.

Key words: Boyer–Moore theorem prover, Nqthm, automated reasoning, interaction.

1. Introduction

Numerous papers have been written to describe challenge problems solved by using the Boyer–Moore theorem prover, Nqthm[1] [3, 6].[2] There are also papers on new tools and functionalities added to Nqthm (for example, first-order quantification [18]) and on applications, including an entire issue of the *Journal of Automated Reasoning* [2]. However, the literature about this system lacks a tutorial paper that focuses on how to use it to solve hard problems and what kinds of issues arise in these situations.[3]

* This research was supported in part by ONR Contract N00014-94-C-0193. The views and conclusions contained in this document are those of the author(s) and should not be interpreted as representing the official policies, either expressed or implied, of Computational Logic, Inc., the Office of Naval Research, or the U.S. government.

In this paper we give a tutorial introduction to Nqthm by illustrating how we have used this system on an example suggested at an inductive theorem-proving workshop [8], the Arithmetic–Geometric Mean theorem. Informally, the theorem says that the arithmetic mean is greater than or equal to the geometric mean for nonempty sequences of nonnegative real numbers:

$$\left(\sum_{i=1}^{n} x_i\right)/n \geqslant \left(\prod_{i=1}^{n} x_i\right)^{(1/n)}.$$

However, the formalization we have chosen to take is the following, which makes sense over the natural numbers as well:

$$\left(\sum_{i=1}^{n} x_i\right)^n \geqslant n^n * \prod_{i=1}^{n} x_i,$$

where $x_i \in \mathcal{N}$ for $0 < i \leqslant n$. Notice that this is a fairly interesting theorem, not completely obvious in our view, but we stress that our purpose here is not to provide the simplest proof of it. In fact, there are simpler proofs over the nonnegative real (or rational) numbers, and perhaps over the natural numbers as well (although we are not familiar with any). The purpose of this paper is not presentation of the mathematics *per se*, but rather a description of the process of *mechanization* of that mathematics through Nqthm.

We would like to think that the paper is accessible to readers without knowledge of Nqthm. For this purpose we have added enough background material to make the document self-contained (there are only few 'advanced' details in the presentation that require pointers to other references). In [21] we provide two layers of appendixes that provide many more details, together with a complete input file for Nqthm leading to the proof of the main theorem.

We believe that it is important to provide a *user* view of interaction with theorem provers such as Nqthm, for two reasons.

- A user view of a theorem prover is useful, of course, for prospective *users* of this kind of technology.
- A user view of a theorem prover is useful as a guide to *implementors* of theorem provers who are looking for ways to make their systems more practical to use.

To present a user's perspective on this system, we have tried to give an honest account of what transpired in the sessions we had with Nqthm, including some of our missteps. We do clean things up just a bit when that significantly improves the exposition, such as introducing the `LET` construct for readability (though we *could* have done so in the first place!). To support our focus on user-level perspective, we do not always give an analysis of what the prover did, at least not to the extent that a researcher in automated reasoning might be tempted to give. (Readers who want more of an implementation-level perspective on Nqthm are referred to [3].) Notice that we *can* get away from the implementor's perspective

– that is, we can avoid looking too much inside what the prover does – because Nqthm gives feedback that is meaningful to *users*, **not just to implementors**. The implementor's perspective may be much more common in the automated reasoning literature.

As suggested above, we hope that this work will be valuable both for Nqthm users and implementors of automated deduction systems. Nqthm users will find in this document an extensive panorama, by means of examples, of the approaches and the tricks that can be used to prove theorems with Nqthm. Implementors could find interesting the view of interaction with Nqthm imparted by this 'empirical' work, especially as highlighted in Section 3.

Note that this paper concerns the *interactive* use of Nqthm, which is the most common style of use. We do not address uses where Nqthm is required to succeed totally automatically (see, for example, [4] and [19]).

1.1. Structure of the Paper

In Section 2 we give some background material about Nqthm, its 'interactive enhancement' Pc-Nqthm,[4] and the 'checkpoint' tool. In Section 3 we make some observations about the kind of interaction that an (accomplished) user has with Nqthm. Section 4 contains the hand proof that guided our mechanized proof. Section 5 is devoted to the description of our mechanized proof, in the spirit of an annotated transcript. Finally, Section 6 contains some concluding remarks.

1.2. Notational Conventions

To guarantee the tutorial flavor of this paper, we have decided to present Nqthm expressions maintaining their Lisp (s-expression) form (see Subsection 2.1), so that the user need not guess what we actually did with the system. Text appearing on a line after a semicolon is a comment.

In our presentation, the prover input commands that end up in our final script are displayed in UPPER CASE; they are in lower case when they fail or when they are later undone.

Each time we first introduce an Nqthm command in the proof (Section 5), we explain the syntax and meaning of that command.

1.3. Acknowledgments

We thank Fausto Giunchiglia and Rich Cohen for comments on drafts of this paper. In addition, we are grateful to the referees for a number of useful comments. We also thank Alan Bundy and Deepak Kapur for suggesting that we take on this example.

Matt Kaufmann thanks the Mechanized Reasoning Group of IRST (Povo, Italy), in particular its head Fausto Giunchiglia, for their hospitality during several days in the summer of 1994, during which time the proof in this paper was performed.

And, of course, none of this work would be possible without the Nqthm theorem prover of Bob Boyer and J Moore, who also have been primary in the formulation of many of the ideas presented in this tutorial.

Mahadevan Subramaniam previously proved a version of the Arithmetic–Geometric Mean theorem using the RRL theorem prover. His success encouraged us to make an attempt.

2. Background on Nqthm, Pc-Nqthm, and Checkpoints

The first and the third subsections below describe the Nqthm logic and theorem prover, respectively. They are taken with permission, with some small changes and additions, from [9], which in turn probably took some of its words from the Nqthm 'folklore'. For further details see [6]. The words in the Subsection 2.5 are adapted from [13]. See also [7] for more about Nqthm and Pc-Nqthm and about applications of these systems, and for a somewhat simpler example of their use than the one presented here.

2.1. The Nqthm Logic

The logic of Nqthm (sometimes called the 'Boyer–Moore Logic') is a simple quantifier-free fragment of first-order logic resembling in syntax and semantics the Lisp programming language. Terms in the logic are written using a prefix syntax, which is case-insensitive – we write `(PLUS I J)` where others might write `PLUS(I,J)` or `i+j`. The logic is formally defined as an extension of propositional calculus with variables, function symbols, and the equality relation. Axioms are added defining the following:

- the Boolean (logical) constants `(TRUE)` and `(FALSE)`, abbreviated `T` and `F`;
- the if-then-else function, `IF`, with the property that `(IF x y z)` is `z` if `x` is `F` and `y` otherwise;
- the Boolean connectives `AND`, `OR`, `NOT`, and `IMPLIES`;
- the equality function `EQUAL`, with the property that `(EQUAL x y)` is `T` or `F` according to whether `x` is `y`;
- and inductively constructed objects including natural numbers and ordered pairs (which also represent lists):
  - Natural numbers, generated by successor function `ADD1`, with corresponding predecessor function `SUB1`, and *base object* `(ZERO)`;
  - Lists and trees, generated by ordered-pair constructor `CONS` and accessors `CAR` and `CDR`, and recognized by the predicate `LISTP`.[5] When such objects are viewed as lists, `CAR` returns the first element of any non-empty list, `CDR` returns the tail of such a list, and `NIL` is an atom typically used

to represent the empty list. (These names are borrowed from the family of Lisp languages.)

In addition, there is the ability within the logic to add user-defined inductive data structures, though we do not take advantage of this capability in the arithmetic–geometric mean example.

Nqthm offers a construct, LET, for introducing 'local variables'. The syntax is probably best illustrated by an example, such as the following:

```
(LET ((X (FOO X))
      (Y (G U V)))
     (H X Y U))
```

is an abbreviation for the expression

```
(H (FOO X) (G U V) U)
```

The logic also provides a principle of recursive definition under which new function symbols may be introduced. The following, for example, is a definition within the logic of a list concatenation function APPEND.

```
(APPEND X Y) = (IF (LISTP X)
                   (CONS (CAR X) (APPEND (CDR X) Y))
                 Y).
```

This equation submitted as a definition is accepted as a new axiom under certain conditions that guarantee that one and only one function satisfies the equation. One of these conditions is that certain derived formulas be theorems. Intuitively, these formulas ensure that the recursion terminates by exhibiting a 'measure' of the arguments that decreases, in a well-founded sense, in each recursive call of the function. Several functions are defined in the initial 'ground-zero' theory, including standard recursive definitions of the arithmetic functions PLUS and TIMES over the natural numbers, and the 'less-than' relation on the natural numbers, LESSP with its cousins LEQ (less than or equal) and GEQ (greater than or equal).

The rules of inference of the logic, in addition to those of propositional calculus and equality, include mathematical induction. The formulation of the induction principle is similar to that of the definitional principle. To justify an induction scheme it is necessary to prove certain theorems that establish that, under a given measure, the inductive hypotheses are about 'smaller' objects than the conclusion. Using induction, one can prove such theorems as the associativity of the APPEND function defined above; this can be stated as a theorem in the logic.

```
Theorem ASSOCIATIVITY-OF-APPEND
(EQUAL (APPEND (APPEND A B) C)
       (APPEND A (APPEND B C)))
```

Notice that this theorem provides a partial *specification* of the APPEND function. It is one of myriad properties of this function and its relation to others that can be defined and proved within the logic.

2.2. Brief Introduction to Conditional Rewriting

The theorem above, once proven, may be used in the proofs of subsequent theorems if it is labeled as a *rewrite rule* (syntax presented in the next subsection). That is, we can interpret this theorem as sanctioning the replacement of any expression that matches the left-hand side of the equality by the corresponding instance of the right-hand side. Thus, for example, the expression

```
(*)         (APPEND (APPEND V (APPEND W X)) (APPEND Y Z))
```

can be readily shown to be equal to

```
(**)        (APPEND V (APPEND W (APPEND X (APPEND Y Z))))
```

by two applications of ASSOCIATIVITY-OF-APPEND interpreted as a rewrite rule. By defining and proving rewrite rules, it is possible to build up a powerful theory for proving interesting facts about a specific domain. In fact, probably the most important activity of most users when interacting with Nqthm is the formulation of rewrite rules, which will be applied automatically during subsequent proof attempts. For example, if the prover sees the term (*) above during a proof, and if the theorem ASSOCIATIVITY-OF-APPEND has already been proved and stored as a rewrite rule, then the prover will replace (*) with (**).

Rewrite rules in Nqthm may be *conditional*; that is, they may have hypotheses. Consider for the sake of example the following somewhat weaker version of the rewrite rule above, which says that the variable A must be an ordered pair (a nonempty list):

```
Theorem ASSOCIATIVITY-OF-APPEND
(LISTP A)
 -->
(EQUAL (APPEND (APPEND A B) C)
       (APPEND A (APPEND B C)))
```

Then to apply this lemma to the same term as above,

```
(APPEND (APPEND V (APPEND W X)) (APPEND Y Z)),
```

the Nqthm rewriter would call itself recursively, attempting to prove

```
(LISTP (APPEND V (APPEND W X))).
```

If that recursive rewrite were to fail, then the rewriter would not apply this lemma to this term.

(Note: Unlike some other provers, Nqthm does not implement any method for orienting rewrite rules or proving termination of sets of rewrite rules. The rewriter is described in [3]. Although it contains some subtleties, the main thing to understand about it is that it uses rewrite rules and non-recursive definitions in the manner described above, and is somewhat more cautious in using recursive definitions.)

Next we take a look at the Nqthm theorem prover as a whole, keeping in mind that (conditional) rewriting tends to be the most important part of the activity of this system.

2.3. The Nqthm Theorem Prover

The Nqthm theorem prover is a computer program that takes as input a conjecture formalized as a term in the logic and attempts to prove it by repeatedly transforming and simplifying it. The theorem prover employs the following basic transformations:

- *simplification*, which includes
  - decision procedures for propositional calculus, equality, and linear arithmetic,
  - conditional rewriting based on axioms, definitions, and previously proved lemmas, and
  - automatic application of user-supplied simplification procedures that have been proven correct;
- elimination of calls to certain 'destructor' functions in favor of others that are 'better' from a proof perspective;
- heuristic use of equality hypotheses;
- generalization by the replacement of terms by variables;
- elimination of apparently irrelevant hypotheses; and
- mathematical induction.

The theorem prover contains many heuristics to control the orchestration of these basic techniques. However, as we have already suggested, it is rarely necessary or even useful for the user to have deep knowledge of such heuristics.

Definitions are submitted to the system by using the DEFN command. The syntax of definitions in Nqthm is as follows:

```
(DEFN name arguments body)
```

For example, the syntax used to define the APPEND function, introduced in Subsection 2.1, is as follows.

```
(DEFN APPEND (X Y)
  (IF (LISTP X)
      (CONS (CAR X) (APPEND (CDR X) Y))
      Y))
```

Later we will see that an additional argument may be supplied, which can be used to 'explain' why a given recursively defined function terminates.

The system displays a script of the proof attempt, allowing the user to follow the progress of the proof and take steps to abort misdirected proof attempts. From the script it is often apparent to the skilled user how to improve the prover's knowledge base so that a subsequent proof attempt will succeed. The script printed by the prover in discovering the proof of the lemma named `ASSOCIATIVITY-OF-APPEND` is shown in Figure 1. That lemma follows the syntax of theorems in Nqthm:

```
(PROVE-LEMMA name rule-classes formula)
```

where `name` is a name to be associated with the theorem, `formula` is the theorem to be proved, and `rule-classes` indicates how the theorem is to be stored. Typically `rule-classes` is `(REWRITE)`, meaning that the theorem should be stored as a rewrite rule, in the sense described in Subsection 2.2. We will also see on occasion an extra, optional argument, which is a list of hints to the theorem prover.

In a shallow sense, the prover is fully automatic; the system accepts no advice or directives from the user once a proof attempt has started. The only way the user can alter the behavior of the system during a proof attempt is to abort the attempt. However, in a deeper sense, the theorem prover is interactive; the system's behavior is influenced by the database of lemmas that have already been formulated by the user and proved by the system. Each conjecture, once proved, is converted into one or more rules and stored in the prover's database to guide the theorem prover's actions in subsequent proof attempts. Often these are rewrite rules, but there are other classes of rules as well. A database is thus more than a logical theory; it is a set of rules for proving theorems in the given theory. We consider interaction in Section 3.

Using this approach, the Boyer–Moore prover has been used to check the proofs of some quite deep theorems. See [7, 6], or [13] for rather lengthy lists of applications of the system (with references).

We conclude our description of the Nqthm prover by pointing out that at a coarse level, interaction with Nqthm proceeds at the level of *events*. An event is the introduction of a definition (such as `APPEND`), the statement of a theorem (such as the `PROVE-LEMMA` form above), or any of a few other *event forms*. The user's goal is to formulate appropriate definitions and theorems and create a file of events that includes not only those, but also any additional events necessary to support the proofs for the desired events. At all points during a session, the Nqthm database contains a stack of events that have already been processed; each time it successfully processes a new event, it pushes the new event on the stack, which is called a *history*. There are a few other commands that are useful besides events. For example, the command `(UBT)`, 'undo-back-through', pops an event off the history stack; when given a name as an argument, it pops all

```
(PROVE-LEMMA ASSOCIATIVITY-OF-APPEND (REWRITE)
        (EQUAL (APPEND (APPEND X Y) Z)
               (APPEND X (APPEND Y Z))))

        Call the conjecture *1.

        Perhaps we can prove it by induction.  Three inductions
    are suggested by terms in the conjecture.  They merge into
    two likely candidate inductions.  However, only one is
    unflawed. We will induct according to the following scheme:

        (AND (IMPLIES (AND (LISTP X) (p (CDR X) Y Z))
                      (p X Y Z))
             (IMPLIES (NOT (LISTP X)) (p X Y Z))).

    Linear arithmetic and the lemma CDR-LESSP can be used to
    prove that the measure (COUNT X) decreases according to the
    well-founded relation LESSP in each induction step of the
    scheme. The above induction scheme leads to two new goals:

    Case 2. (IMPLIES (AND (LISTP X)
                          (EQUAL (APPEND (APPEND (CDR X) Y) Z)
                                 (APPEND (CDR X) (APPEND Y Z))))
                     (EQUAL (APPEND (APPEND X Y) Z)
                            (APPEND X (APPEND Y Z)))),

    which simplifies, applying the lemmas CDR-CONS and CAR-CONS,
    and opening up the definition of APPEND, to:

          T.

    Case 1. (IMPLIES (NOT (LISTP X))
                     (EQUAL (APPEND (APPEND X Y) Z)
                            (APPEND X (APPEND Y Z)))),

    which simplifies, unfolding the function APPEND, to:

          T.

       That finishes the proof of *1.  Q.E.D.
```

Figure 1. Proof of the lemma ASSOCIATIVITY-OF-APPEND.

events back through that name, inclusive. Other Nqthm commands that we use are explained below at the first time they are used.

2.4. Introduction to Induction Schemes in Nqthm

We turn now to a very informal introduction to the relation in Nqthm between recursive function definitions and induction schemes. By adopting a very informal style, we hope to make the presentation below self-contained for all but the most extreme logical purists (who probably would be able to work out details of the Nqthm logic from this informal description, anyhow!).

In traditional proofs by (strong) mathematical induction, or more precisely *Noetherian* induction, if one wants to prove a proposition of the form $\forall \vec{x} P(\vec{x})$, one may assume $P(\vec{y})$ for all $\vec{y}$ that are 'smaller' than $\vec{x}$ in an appropriate sense. That is, it suffices to prove a statement of the form "(for all $\vec{y}$ smaller than $\vec{x}$, $P(\vec{y})$) **implies** (for all $\vec{x}$, $P(\vec{x})$)". However, the Nqthm logic does not have quantifiers; in particular, one cannot express the preceding statement in that logic. Instead, that logic has an induction principle (stated precisely in Chapter 4 of [6]) that serves much the same purpose as the Noetherian induction principle stated informally above. Here is an overview that should suffice for the purposes of this paper; for more details, see [6].

The Nqthm idea of inductive proof is based on the idea of breaking a proof into cases and then assuming instances of the theorem as inductive hypotheses in a given case, as long as these instances replace variables with terms that are appropriately 'smaller' under the assumption of that case. Significantly, this decomposition into cases, each with zero or more induction hypotheses, corresponds to the structure of recursive definitions in Nqthm. Let us illustrate this correspondence by way of a simple example: a definition of a function that returns the length of a list, which we will need later anyhow.

```
(DEFN LENGTH (X)
  (IF (LISTP X)
      (ADD1 (LENGTH (CDR X)))
    0))
```

Suppose now that we want to prove a proposition `p(X)`. Let us break our goal into two cases according to the `IF` structure in the definition above.

Case 1: `(LISTP X)`. Then (if we believe that the definition of `LENGTH` terminates), it 'should' be the case that the `(CDR X)` is 'smaller' than `X` in this case, in some appropriate sense. So when proving `p(X)`, we 'may' assume the inductive hypothesis, which is `p((CDR X))` in this case, in addition to the case hypothesis `(LISTP X)`.

Case 2: otherwise. Then since there are no recursive calls of `LENGTH` in this 'branch' of the `IF` structure of the body of the definition, we make no inductive hypothesis in this case. After all, this case is what we would normally call a 'base case'.

Now *suppose* that we want to prove a proposition by the induction argument above, where we break into cases `(LISTP X)` and `(NOT (LISTP X))`, and where we assume in the former case that the proposition holds with `(CDR X)` replacing `X`. *Then* we can imagine simply saying: "Use the induction scheme suggested by the definition of `LENGTH`." And, in fact, if we were to give to Nqthm the hint

```
(induct (length x))
```

to a PROVE-LEMMA event, we would be telling the theorem prover to prove the indicated theorem by just that induction argument. We can display the structure of that induction argument as follows; see also [3] for a more thorough presentation of this idea.

```
Case 1:   (LISTP X)
  For the inductive hypothesis:  replace X by (CDR X)

Case 2:   (NOT (LISTP X))
  No inductive hypothesis.
```

Note that in order to believe in the soundness of this induction scheme, we need to know that (CDR X) is 'smaller' than X when (LISTP X) holds, where 'smaller' is in the sense of a well-founded relation. In fact there is a function COUNT with the property that in the base Nqthm theory, it is a theorem that (LISTP X) implies that (COUNT (CDR X)) is less than (COUNT X) in the usual ordering of the natural numbers, which is well founded.

In Subsection 5.3 we will construct a more complicated induction scheme than the one shown in the two cases above. One can imagine a syntax for introducing induction schemes, where one supplies the cases and corresponding substitutions (for the inductive hypotheses). In that case there would be a proof obligation that is the same as the 'termination proof' required for a recursive definition with the same case/substitution structure. However, the introduction of such syntax seems an unnecessary complication, since successful Nqthm users already understand the connection between recursive definitions and induction schemes. Such a facility, therefore, seems rather redundant and syntactically noisy.

2.5. The Pc-Nqthm Interactive Enhancement

Pc-Nqthm ('Proof-checker Nqthm') is an interactive enhancement of Nqthm. The system is described in detail in [7, 11, 12], and [18]; here is a brief introduction.

The user enters an interactive loop with the command VERIFY. Consider the example of the associativity of append from the preceding subsection:

```
(verify (equal (append (append x y) z)
               (append x (append y z))))
```

The system replies with a prompt, '->:', indicating that one is now in an interactive loop. In this loop, the user can give commands at a low level (such as deleting a hypothesis, diving to a subterm of the current term, expanding a function call, or applying a rewrite rule) or at a high level (such as invoking the

Nqthm prover). For example, the INDUCT command replaces the current goal by subgoals corresponding to the base and induction steps. In the following transcript, everything except the prompt and 'induct' was printed by the system.

```
->: induct

Inducting according to the scheme:

      (AND (IMPLIES (AND (LISTP X) (p (CDR X) Y Z))
                    (p X Y Z))
           (IMPLIES (NOT (LISTP X)) (p X Y Z)))

Creating 2 new subgoals, (MAIN . 1) and (MAIN . 2).

The proof of the current goal, MAIN, has been completed.  However,
the following subgoals of MAIN remain to be proved: (MAIN . 1) and
(MAIN . 2). Now proving (MAIN . 1).
->:
```

Commands also exist for displaying useful information (such as printing the current hypotheses and conclusion, displaying the currently applicable rewrite rules, or showing the current abbreviations), and controlling the progress of the proof (such as undoing a specified number of commands, changing goals, or disabling certain rewrite rules). A notion of 'macro commands' lets the user create compound commands, roughly in the spirit of the tactics and tacticals of LCF and its descendents. An on-line help facility is provided.

As with a variety of proof-checking systems, this system is goal-directed: a proof is complete when the main goal and all subgoals have been proved. Upon completion of an interactive proof, the lemma may be added to the Nqthm history. This event can later be replayed in 'batch mode'. Partial proofs can also be stored.

See [7] or [13] for a number of applications of Pc-Nqthm (with references).

2.6. Checkpoints

Intensive users of theorem provers often rely on feedback from the system they use when a proof attempt, or proof step, fails to achieve the desired effect. Nqthm users are no exception, and those who are successful tend to be accomplished at making sense out of the theorem prover's output. They seem to have a knack for looking at the output from a failed proof attempt and quickly deciding on an approach that seems promising for guiding the theorem prover to success. How do they do it?

The main trick is to focus on goals that are stable under simplification, i.e., are left unchanged by the Nqthm simplifier, which is described briefly near the beginning of Subsection 2.3. Following (roughly) an idea in [6], we will refer to goals that are stable under simplification as *checkpoints*. Why is it so useful to find goals that are stable under simplification? Very often it is the case that these goals are 'obvious' to the user, and will become 'obvious' to the theorem prover as well if a useful rewrite rule can be proved. For example, a goal such as

```
(implies (and (p x) (p y))
         (equal (foo (append (append x y) (h x y)))
                (foo (append x (append y (h x y))))))
```

seems obvious because APPEND is associative. If this goal is stable under simplification, then apparently the theorem prover simply does not yet 'know' that APPEND is associative. The next step, then, would be to prove such a lemma as a *rewrite rule*, so that it can be used automatically by the simplifier. Such a rule is proved in Subsection 2.3. A new proof attempt could then proceed past the point of the goal displayed above.

During the present proof effort, we employed a tool that we refer to in this paper as the 'checkpoint tool'. This tool, which is documented in [14],[6] allows the user who is running Nqthm under Gnu emacs [20] to get the cursor to move quickly to a checkpoint, which will be a goal that is stable under simplification. Furthermore, the tool selects additional checkpoints upon request, according to a heuristic prioritizing scheme.

3. Interacting with Nqthm

In this section we give a high-level description of (our view of) Nqthm interaction. These points are exemplified in detail, many times over, in Section 5.

In our view, there is no deep mystery to typical interaction with Nqthm. The user formulates theorems and supporting definitions, then invents a main strategy, and finally proves theorems and modifies definitions by inspecting goals from failed proofs that are left unchanged by simplification. (See the description of checkpoints in Subsection 2.6.)

Somewhat more precisely, we can say that at a 'low' (non-strategic) level, the idea is to use Nqthm primarily as a simplifier, though allowing some use of induction and other proof techniques, to see where the simplification 'stalls'. This is very different from our usual informal style of mathematical reasoning, where we try to stay at the 'strategic' level rather than thinking in terms of 'canonicalization'. Sometimes it is best to think at a high level but to interact at that sort of lower level – when trying to wrestle with the prover, it's best not to think strategically, but instead, to think in terms of helping the simplifier by proving useful rewrite rules so that it can simplify further than it did before.

To summarize the two main components of Nqthm interaction:

- Decompose the entire theorem into lemmas that yield it immediately, even before descending into the theorem prover. This decomposition isn't necessarily an obvious syntactic one, as we explain below.
- Submit a given theorem to the prover, inspect goals that are stable under simplification (perhaps using the checkpoint tool; see Subsection 2.6), and

then prove lemmas (especially, rewrite rules) to help the prover with those failed goals.

In addition (fortunately or unfortunately), sometimes we find it helpful to employ subtle 'advanced' techniques in our user interaction. For example, we occasionally use the interactive facility provided by Pc-Nqthm (see Subsection 2.5) for 'debugging', and sometimes we look at a trace of the rewrites that have been performed (see, for example, Subsection 5.4).[7] And in fact, sometimes we test our functions using the Nqthm execution tool, `R-LOOP` (see, for example, Subsection 5.4), perhaps even running tests that show us that our 'theorems' are not theorems after all!

Returning to the technique of providing high-level structure, sometimes it is very useful to think about the underlying mathematics when decomposing theorems into lemmas, in a manner typical of doing mathematics. For example, we proved `MIN + 1 < MAX` (see Subsection 5.6) by thinking about how we would reason informally, namely, `MIN < AVERAGE < MAX`. Then, our lemma structure reflected this informal reasoning.

In conclusion, our view is that support for interaction with Nqthm is pretty good already, *if* one knows what he/she is doing (for example, absorbs this paper). It could presumably be improved, but the improvements are not necessarily in the organization of the prover's processes or heuristics. So we pose the following question.

BIG QUESTION:
How could interaction with Nqthm
(and similar theorem provers) be improved?

Of course, there are no definitive answers to this question. But we believe that in order to make theorem provers useful to an audience wider than the automated reasoning research community, it is important to avoid intertwining interaction and internal prover control.[8] Although the Nqthm user will find it useful to adopt a certain paradigm of interaction, nevertheless we do *not* think that deep understanding of the theorem prover's internal workings is particularly useful; in fact, it probably confuses the matter more than it helps. Unfortunately there are occasional tricky points; for example, in the course of the proof presented below, the prover (more precisely, the linear arithmetic decider) is not able to infer that `MAX` is a natural number from the fact that `MIN < MAX` (see Subsection 5.10). So maybe the answer to the 'big question' above is: improve the prover to get rid of such rough edges, so that the user is indeed shielded from low-level prover implementation issues. Notice that a *user* is not necessarily an *implementor*! These two classes of individuals have different concerns. So, for example, if the prover is not behaving as expected, we take the position that a reasonable user response is "How do I overcome or work around this problem?", while the implementor's response is more likely something like "Why didn't the prover do a better job here?” In other words, we are taking the position that the *user* would prefer not to be bothered with such ‘why’ questions. We will demonstrate below that, in the case of Nqthm, one can actually profit from such an attitude.

It is beyond the scope of this paper to make comparisons between Nqthm and other theorem proving systems. We suspect that the particular proof presented in the next section is nontrivial to ‘code’ into any existing theorem prover without a certain amount of low level interaction, even provers that attempt to focus on a high (‘strategic’) level of interaction. *Detailed* expositions that present logs of successful proofs but stay at a high level would therefore be of interest, at least to us.

4. The Hand Proof

In this section we give a short hand proof of a formalization of the arithmetic–geometric mean theorem from Section 1. It is easy to imagine a much more detailed hand proof; for example, several observations require the use of mathematical induction for a rigorous proof. Thus, the contrast in length between this hand proof and our mechanized proof is due not only to the cost of mechanization, but can be viewed in part as the cost of providing ‘reasonable’ details. That said, we choose to present a short proof in this section, omitting details that might distract one from the main ideas.

In the following, $\vec{x} = \langle x_1, \ldots, x_n \rangle$ denotes a nonempty sequence of natural numbers. Moreover, we assume that max and min are functions returning respectively the maximum and the minimum numbers contained in a sequence of natural numbers.

LEMMA 1. *If* $\max(\vec{x}) \neq \min(\vec{x})$ *and there exists a natural number* k *such that*

$$\sum_{i=1}^{n} x_i = n * k,$$

then

$$\min(\vec{x}) + 1 < \max(\vec{x}).$$

Proof. Notice that the hypothesis implies that k is the average of the elements contained in $\vec{x}$. Therefore, from the assumption that the ‘maximum’ and the ‘minimum’ of $\vec{x}$ are different, we have that

$$\min(\vec{x}) < k < \max(\vec{x}).$$

□

The following lemma may be viewed as a special case of our goal, where the average is an integer.

LEMMA 2. *If there exists a natural number k such that*

$$\sum_{i=1}^{n} x_i = n * k,$$

then

$$k^n \geqslant \prod_{i=1}^{n} x_i.$$

Proof. By well-founded induction on the lexicographic relation $\prec$ defined as follows

$$\vec{x} \prec \vec{y} \equiv (\max(\vec{x}) < \max(\vec{y}) \vee (\max(\vec{x}) = \max(\vec{y}) \wedge occ(\max(\vec{x}), \vec{x}) < occ(\max(\vec{y}), \vec{y}))),$$

where $occ(a, x)$ equals the number of occurrences of a in the sequence x. It can be easily proved that $\prec$ is well-founded.

We can split the proof in two cases. Very informally, the idea in the 'inductive step' (Case 2), where the maximum and minimum are distinct, is to decrease the maximum by 1 and increase the minimum by 1, which does not change the average of the list but can only increase its product (hence, its geometric mean). Then, the inductive hypothesis applies.

Case 1: $\max(\vec{x}) = \min(\vec{x})$. It follows that $k = x_i$ for each i, $1 \leqslant i \leqslant n$, and hence

$$k^n = \prod_{i=1}^{n} x_i.$$

Case 2: $\max(\vec{x}) \neq \min(\vec{x})$. Then we may choose distinct k and j such that x_k, x_j are respectively the maximum and the minimum of $\vec{x}$. Consider the sequence $\vec{y} = \langle y_1, \ldots, y_n \rangle$ obtained from $\vec{x}$ by replacing x_k with $x_k - 1$ and x_j with $x_j + 1$. Note that the sums of the elements respectively in $\vec{x}$ and $\vec{y}$ are equal. Moreover, we have

$$\prod_{i=1}^{n} y_i > \prod_{i=1}^{n} x_i, \tag{1}$$

since from Lemma 1 and simple algebra we can easily prove that $(x_k - 1) * (x_j + 1) > x_k * x_j$. But now the conclusion follows from (1) together with the inductive hypothesis

$$k^n \geqslant \prod_{i=1}^{n} y_i$$

– once we show that the inductive hypothesis is applicable. That is, it remains to prove $\vec{y} \prec \vec{x}$, which we do in the following two cases, which are presented here in no particular order.

Case 2.1: $occ(\max(\vec{x}), \vec{x}) = 1$. From Lemma 1 it follows that $\max(\vec{y}) < \max(\vec{x})$. Thus $\vec{y} \prec \vec{x}$.

Case 2.2: $occ(\max(\vec{x}), \vec{x}) > 1$. In this case the maxima of $\vec{y}$ and $\vec{x}$ are equal and from Lemma 1 we have that $occ(\max(\vec{y}), \vec{y}) < occ(\max(\vec{x}), \vec{x})$. It follows that $\vec{y} \prec \vec{x}$. □

THEOREM 1.

$$\left(\sum_{i=1}^{n} x_i\right)^n \geqslant n^n * \prod_{i=1}^{n} x_i.$$

Proof. Let $y_i = n * x_i$ for each i, $1 \leqslant i \leqslant n$. We have that

$$\left(\sum_{i=1}^{n} y_i\right) = n * \left(\sum_{i=1}^{n} x_i\right). \tag{2}$$

By applying the previous lemma (with $k = \sum_{i=1}^{n} x_i$ and replacing $\vec{x}$ by $\vec{y}$), we obtain

$$\left(\sum_{i=1}^{n} x_i\right)^n \geqslant \prod_{i=1}^{n} y_i = \prod_{i=1}^{n} (n * x_i) = n^n * \prod_{i=1}^{n} x_i. \tag{3}$$

□

5. A High-Level View of the Mechanized Proof

To make our presentation of the mechanized proof more readable, we have divided it into subsections. Our presentation roughly parallels the hand proof in Section 4.

- In Subsection 5.1 we start the Nqthm session defining the sum and the product of a sequence (actually a list) of numbers, and we state the theorem to prove.
- In Subsection 5.2, we prove the theorem assuming as an axiom an Nqthm term, which we call the main lemma, corresponding to Lemma 2 in the hand proof.
- At this point we turn to the task of proving the main lemma by induction. Subsection 5.3 contains the formalization of the induction argument by way of an Nqthm definition.
- In Subsection 5.4 we describe our first attempts to admit this definition. These attempts culminate in the discovery of a bug in our first definition of a function that deletes one occurrence of an element from a list. Here we

do not prove any lemma that is relevant for the final script. Nevertheless we take this digression, not only because it faithfully represents the order in which we carried out the various parts of the proof, but more importantly, because it illustrates a combination of important techniques in the use of Nqthm.

- In Subsection 5.5 we outline the termination proof for the definition that formalizes the induction argument, splitting into two cases.
- Next, in Subsection 5.6, we take a break from the main flow of the proof and prove a version of Lemma 1 from the hand proof. We really did take this break; in fact, this kind of 'jumping around' is not atypical of user interaction with Nqthm.
- Next, in Subsection 5.7, we prove one of the two main cases for termination, corresponding to Case 2.2 from the hand proof in Section 4.
- In Subsection 5.8, we prove the other case for termination, Case 2.1 from the hand proof.
- Finally, we conclude in Subsection 5.9 that the definition we have been considering, which formalizes the induction argument, actually terminates.
- Subsection 5.10 contains the 'rest' of the proof; that is, we prove the main lemma, and we use it to prove the desired theorem as done in Subsection 5.2. The proof of the main lemma is divided in two parts: a 'base' case in which we assume that the maximum and the minimum of the sequence are equal, and an 'induction step' in which we assume that they are different. They correspond to Case 1 and Case 2 in the hand proof of Lemma 2.
- Subsection 5.11 contains a more elegant form of the theorem obtained using the Nqthm iteration construct `FOR`.

Notice that we are approaching the proof in a top-down fashion that corresponds to the top-down structure of the proof presented in the preceding section. That is, we reduce the theorem to Lemma 2, then start the proof of Lemma 2, and finally prove Lemma 1 when necessary. Some of the particulars of the proof may appear different from the preceding section, but the same top-down outline drives them both.

5.1. FORMALIZING THE PROBLEM

We believe that it's reasonable to build on existing work when trying to do serious proof development. We therefore begin by taking advantage of an existing library including definitions and proved theorems about natural numbers that others have found useful.[9] The `NOTE-LIB` command instructs Nqthm to include all the facts in this library; the optional second argument `T` instructs Nqthm to load a corresponding compiled file, which can be useful in the execution of Lisp functions that correspond to Nqthm definitions.

```
(NOTE-LIB ''naturals'' T)
```

Perhaps we should add that the first form in the ``naturals'' library is (NOTE-LIB ``bags'' T); that is, the ``naturals'' library includes a library of facts and definitions about bags (multisets).

Let us now define a function that returns the product of the members of a given list. (See Subsection 2.3 for an introduction to Nqthm definitions.) The following recursive definition says that the product of the members of a non-empty list is the result of multiplying the first member (its CAR) by the product of the rest of its elements (the CDR); and, the product of the empty list is 1.

```
(DEFN PRODLIST (LST)
  (IF (LISTP LST)
      (TIMES (CAR LST)
             (PRODLIST (CDR LST)))
    1))
```

A function SUMLIST may be defined entirely analogously, which returns the sum of the members of a list.

Is it really necessary to write such definitions recursively? In Nqthm there is a 'quantifier' construct, FOR, which is actually an iterator that can be useful in defining functions such as the PRODLIST and SUMLIST above. However, our experience is that the meta-level machinery needed to support reasoning about FOR is awkward; it seems much simpler to quickly type in a primitive recursive definition, as we do here. (However, see also Section 5.11.)

Let us state what we are trying to prove. Recall our convention from Subsection 1.2: since the following event is displayed in lower case, Nqthm must not yet be able to prove this theorem without further help. We will provide such help in subsequent subsections. The syntax of PROVE-LEMMA events is explained in Subsection 2.3. In particular, the notation '()' following 'main' below indicates that we do not intend to make a (rewrite or any other) rule out of this theorem. Also, here EXP is the exponentiation function for natural numbers, which is defined in the ``naturals'' library that we started with. Indeed, Nqthm employs a logic of total functions, and hence GEQ must 'make sense' on arbitrary Nqthm objects (even Booleans and lists), not just numbers. The Nqthm arithmetic primitives all have the property that they treat values that are not natural numbers exactly as though they were 0. Now, finally, here is a formalization of what we are trying to prove in this tutorial.

```
(prove-lemma main ()
             (let ((n (length a)))
               (geq (exp (sumlist a) n)
                    (times (exp n n) (prodlist a)))))
```

5.2. A Top-Down Approach

Sometimes it is useful to proceed in a top-down fashion. Thus, let us assume the following 'main lemma' as an axiom for now; it corresponds to Lemma 2 in the hand proof of the theorem. Then, we will try to derive the theorem from it. Once we succeed in doing so, we will back up and formulate events that culminate in the proof of the main lemma; and *then*, we will paste down the events that we constructed to prove the theorem from the main lemma.

Here, then, is our main lemma. We choose not to label this as a rewrite rule; we will see below how to explicitly direct the use of a specified instance of this lemma when proving the main theorem. The syntax of ADD-AXIOM events is analogous to the syntax of PROVE-LEMMA events.

```
(add-axiom main-lemma ()
           (let ((n (length x)))
             (implies (equal (sumlist x) (times k n))
                      (not (lessp (exp k n) (prodlist x))))))
```

From the hand proof of the theorem, we can see how this lemma proves MAIN, using two trivial facts about scalar product, used informally in Equations 2 and 3 in Section 4:

```
(defn scalar-product (scalar lst)
  (if (listp lst)
      (cons (times scalar (car lst))
            (scalar-product scalar (cdr lst)))
    nil))

(prove-lemma sumlist-scalar-product (rewrite)
             (equal (sumlist (scalar-product scalar lst))
                    (times scalar (sumlist lst))))

(prove-lemma prodlist-scalar-product (rewrite)
             (equal (prodlist (scalar-product scalar lst))
                    (times (exp scalar (length lst))
                           (prodlist lst))))
```

These prove easily. Now we try to prove MAIN from the lemmas above, and after some initial mistakes, we obtain the following formulation.

```
(prove-lemma main ()
             (let ((n (length a)))
               (geq (exp (sumlist a) n)
                    (times (exp n n) (prodlist a))))
             ((use (main-lemma (k (sumlist a))
                   (x (scalar-product (length a) a))))))
```

The last two lines above contain an example of a USE hint. They tell the prover that instead of proving the theorem at hand – let us call it p for the moment – the prover should instead prove (IMPLIES hyp p), where hyp is the result of applying the indicated substitution,

```
{k <- (sumlist a),
 x <- (scalar-product (length a) a)},
```

to the previously-proved theorem called MAIN-LEMMA.

The proof succeeds. So, let's undo the events from this subsection, using the UBT command (see Subsection 2.3), and go back and prove the lemma.

```
(ubt main-lemma)
```

Where are we? Nqthm provides a command for seeing the names of any number of the most recent events in the *history* (see Subsection 2.3): CH. The syntax is

```
(CH n)
```

where n is a natural number and the result is to print out the names of the most recent n events of the history.[10] Now, if we execute (CH 6), say, we can see that the most recent events are the definitions of PRODLIST, SUMLIST, and LENGTH, preceded by events from our initial NOTE-LIB command. Good: that is where we want to be.

5.3. Formalizing the Induction Scheme

Our goal now is to prove the main lemma, so we need to formalize the induction scheme that we have written on paper. Before we begin to do so, we need to define a few functions.

The function MAX happens to be built into Nqthm (and shortly we will display its definition); it returns the maximum of two natural numbers, where arguments that are not natural numbers are treated as zero. In the definition below, for the case that X has only one element, we apply the built-in function FIX, where (FIX X) is equal to X unless X is not a natural number, in which case (FIX X) is 0. Originally we forgot to wrap the call of FIX around (CAR X) in this definition, but during a proof that came later, we found using the checkpoint tool that we needed to prove (NUMBERP (MAXLIST X)) and could not do so. We then inspected the definition of MAXLIST and realized that we had to apply FIX as shown.

```
(DEFN MAXLIST (X)
  (IF (LISTP X)
      (IF (LISTP (CDR X))
          ;; So, X has at least two elements.
          (MAX (CAR X) (MAXLIST (CDR X)))
        ;; Otherwise, X has only one element.
        (FIX (CAR X)))
    ;; We ''shouldn't'' really reach the case where X is empty,
    ;; since we
    ;; terminate above when X has one element.  But anyhow, to
    ;; complete
    ;; the definition in a way that appears reasonable somehow:
    0))
```

MAX is indeed built in, as we see now using the PPE command ('Pretty-Print Event'), which causes the theorem prover to print out the requested event. The last part of the system's response means that MAX is built in.

```
>(ppe 'max)  ; This line is the input; system output follows.

(DEFN MAX
      (X Y)
      (IF (LESSP X Y) Y (FIX X)))

(***** MAX is a satellite of
       (BOOT-STRAP NQTHM))
```

It turns out that we need an analogous notion of minimum, but MIN is not built in. So, let us define MIN analogously. Note that we expect that, for example (treating non-numbers again as 0),

```
(MIN T 3) = (MIN 3 T) = 0,
```

so we should not be surprised at two uses of FIX below.

```
(DEFN MIN (X Y)
  (IF (LESSP X Y) (FIX X) (FIX Y)))
```

And now we can use MIN to define a function returning the minimum of a list.

```
(DEFN MINLIST (X)
  (IF (LISTP X)
      (IF (LISTP (CDR X))
          (MIN (CAR X) (MINLIST (CDR X)))
        (FIX (CAR X)))
    0))
```

We want to make the following definition, but in fact (modulo variable names) this definition of OCCURRENCES turns out to be part of the ''naturals'' library that we included originally (or more precisely, part of the ''bags'' library that is included in the ''naturals'' library).

```
(defn occurrences (elt x)
  (if (listp x)
      (if (equal elt (car x))
          (add1 (occurrences elt (cdr x)))
        (occurrences elt (cdr x)))
    0))
```

Our goal now is to prove the main lemma. We want to use the sort of induction argument described in Subsection 2.4, although the particular induction scheme differs from the one provided by LENGTH. Recall from our hand proof (Section 4) that the idea is to have two cases, according to whether or not the maximum and minimum of the list are equal. Moreover, in the case that they are not equal, we want to assume as an inductive hypothesis the claim that the theorem holds when we replace the list by a new list in which the maximum has been decreased by 1 and the minimum has been increased by 1. Let us call that new list X0, just for the time being.

```
CASE 1:   (EQUAL (MINLIST X) (MAXLIST X))
  This is a ''base case'':  we have no inductive hypothesis for
  this case.

CASE 2:   (NOT (EQUAL (MINLIST X) (MAXLIST X)))
  In this case, we would like to assume as an inductive hypothesis
  that the theorem holds with X replaced by the list X0 described
  above.
```

Recall, however, that we need to know that this induction scheme is 'sound'. Just as Nqthm must accept the definition of LENGTH before it uses the induction scheme corresponding to that definition (see Subsection 2.4), Nqthm must accept an appropriate definition before we can use the induction scheme that we are currently developing. The Nqthm principle of definition requires that X0 be 'smaller than' X in an appropriate sense, assuming (NOT (EQUAL (MINLIST X) (MAXLIST X))). In fact, if we look back at the hand proof, we see that we had some additional information available, namely, that (SUMLIST X) is equal to (TIMES K (LENGTH X)). So, let us revise our induction scheme accordingly.

```
CASE 0:   (NOT (EQUAL (SUMLIST X) (TIMES K (LENGTH X))))
  This is a ''base case''; in fact, the main lemma has the negation
  of this case as a hypothesis, so this case will be trivial.

CASE 1:   (EQUAL (MINLIST X) (MAXLIST X))
          and also (EQUAL (SUMLIST X) (TIMES K (LENGTH X)))
  This is a ''base case'':  we have no inductive hypothesis for
  this case.

CASE 2:   (NOT (EQUAL (MINLIST X) (MAXLIST X)))
          and also (EQUAL (SUMLIST X) (TIMES K (LENGTH X)))
  In this case, we would like to assume as an inductive hypothesis
  that the theorem holds with X replaced by the list X0 described
  above.
```

Here, then, is an Nqthm definition whose induction scheme matches the display above, in the sense discussed in Subsection 2.4. Later we will write a suitable expression in place of the mysterious abbreviation X0 introduced above. Note that we may use any expressions in place of the occurrences of T below, as long as they do not contain a call of INDUCTION-FN, because we are using this definition only to generate the induction argument displayed above.

```
(defn induction-fn (x k)
  (if (not (equal (sumlist x) (times k (length x))))
```

```
      t
    (if (equal (minlist x) (maxlist x))
        t
      (induction-fn X0 k))))
```

Let us turn next to the problem of writing a suitable expression in place of X0 above. That is, we wish to formalize the idea of moving the maximum and minimum towards each other by 1 unit, as described in the informal proof, Section 4. Our intuition is that it will be easier to reason about this notion if we can factor the replacement operation into two apparently simpler operations: deleting an element from a list and CONSing a new element on to the front. That thinking gives rise to the following definition, which fleshes out the template above. It uses a function DELETE1, which deletes one occurrence of an element from a list. We will say more about DELETE1 below, and we will explain the last line of the definition below as well.

```
(defn induction-fn (x k)
  (if (not (equal (sumlist x) (times k (length x))))
      t
    (if (equal (minlist x) (maxlist x))
        t
      (induction-fn (cons (sub1 (maxlist x))
                          (cons (add1 (minlist x))
                                (delete1 (maxlist x)
                                         (delete1 (minlist x) x))))
                    k)))
  ((ord-lessp (cons (add1 (maxlist x))
                    (occurrences (maxlist x) x)))))
```

Alternatively, we could have written recursive functions that look for an element and then decrement (or increment) it ‘in place’; that might well work out fine, but our guess is that our tasks will be simpler if we do the ‘factoring’ indicated above.

The last line of the definition above contains a *measure hint*, which is intended to suggest to Nqthm a reason why it should accept this definition (informally, why the definition terminates). This hint reflects the idea from the hand proof that X0 is ‘smaller than’ X in the sense of the usual lexicographic order on pairs of natural numbers,[11] applied to the pair consisting first of the maximum of X, and then of the number of occurrences of its maximum in X.

The definition above seems to take forever to process, so we interrupt it.[12] It is not surprising that this definition presents a challenge, since the heart of the entire proof is to prove termination of this function. (This is a feature of the mathematics of the approach in the hand proof, not of its Nqthm mechanization.) So, let us focus now on that task.

5.4. First Attempts at the Termination Proof

In this subsection we describe our first attempts to admit the definition of INDUCTION-FN. These attempts culminate in the discovery of a bug in our first definition of the function DELETE1. As mentioned before, we take this digression from the main flow of the proof for pedagogical reasons, since we have found this bug by using a combination of some important techniques and commands of Nqthm.

Here is our first attempt to define a function DELETE1 that deletes one occurrence of an element from a list.

```
(defn delete1 (elt x)
  (if (listp x)
      (if (equal elt (car x))
          ;; Since we are only deleting one occurrence, we do not
          ;; recur.
          (cdr x)
        (delete1 elt (cdr x)))
    ;; Otherwise x is empty; return either x or nil, it doesn't
    ;; matter.
    x))
```

It's not difficult to find a bug in this definition once one is told that there is one; but in our actual session with Nqthm, we are working quickly and not paying much attention!

Our initial attempts at admitting INDUCTION-FN are quite fruitless. In lieu of any better ideas at the moment,[13] we execute the Nqthm command

```
(maintain-rewrite-path t)
```

which tells the prover to monitor the rewriting process. We will resubmit the definition of INDUCTION-FN, then abort the processing of this definition so that we can inspect the *rewrite path* especially to look for looping rewrite rules.

Submitting the definition again causes an error, because one has to execute (UBT) after aborting a DEFN. This is an obnoxious feature of Nqthm, and is the sort of thing that we believe should be minimized in serious user-friendly proof assistants. Fortunately, Nqthm contains very few such idiosyncrasies.

After a while we interrupt the DEFN event and submit the form

```
(break-rewrite)
```

which enters the Nqthm tool for inspecting the *rewrite stack*. This notion is explained amply in [6]; here we restrict our explanation to the particular commands that we invoke within this utility.

Inspection of the next-to-top frame of the rewrite stack yields the following response from the system.

```
---- Frame 1 -----              (persistence 127)
Rewriting the conclusion of the top-level goal:
(EQUAL (ADD1 (MINLIST X)) (MAXLIST X))
under the substitution:
NIL
:
```

This baffles us: why is the prover trying to prove this equality?

Rather than think about that (which seems confusing), we decide to try the definition interactively, using the Pc-Nqthm command `VERIFY-DEFN`, by submitting the form `(DEFN INDUCTION-FN ...)` above but with `VERIFY-DEFN` in place of `DEFN`. The system responds by putting us in the Pc-Nqthm interactive loop, where the goal is exactly that which must be proved in order to accept the definition of `INDUCTION-FN`. We issue the `SPLIT` command, since the goal is a rather complicated conjunction, so that the system will simplify the propositional structure to conjunctive normal form, creating one new subgoal for each resulting conjunct. In fact the system creates two subgoals, the first of which is as follows.

```
(ordinalp (cons (add1 (maxlist x))
                (occurrences (maxlist x) x)))
```

We are able to prove this goal automatically, by issuing the `PROVE` command so that Pc-Nqthm invokes the full power of the Nqthm prover to prove this goal.

When we look at the second goal, we see that there are a number of terms that we might want to prove something about, in order to simplify them. But we can get pretty lost in a sea of such lemmas. Let's think for a moment about our strategy. Recall that we imagine taking two elements out of a list, namely, the maximum and the minimum, and moving them to the front after a decrement and an increment. Let us factor all that into simpler steps: move one element to the front, then move another, then modify the first two elements of the list. So, before thinking too much further, we formulate the following lemma, which may be useful.

```
(prove-lemma maxlist-delete1-rearrange (rewrite)
             (implies (member b x)
                      (equal (maxlist (cons b (delete1 b x)))
                             (maxlist x))))
```

Unfortunately, the proof attempt fails. When we look at the failed proof, where do we look? The proof by induction yields several cases, all but the last of which are proved immediately. One might be tempted to look at the last case, but we believe that it is more useful to look at goals that are stable under simplification, so that attention is not wasted on terms that can be further simplified.

Now, when we look at the failed proof, using the checkpoint tool, we find the following.[14]

```
(implies
 (and (listp x)
      (not (equal (maxlist (cdr x)) (car x)))
      (not (lessp (maxlist (cdr x))
                  (maxlist (delete1 (maxlist (cdr x)) (cdr x)))))
      (member (maxlist (cdr x)) (cdr x))
      (not (lessp (car x) (maxlist (cdr x)))))
 (not (numberp (car x))))
```

But when we stare at this for a while, we realize that there does not seem to be any reasonable argument for the truth of this goal after all, the third hypothesis `(NOT (LESSP ...))` is always true, as is the one following it, and we can make the second and fifth ones true by choosing a list whose `CAR` is bigger than the maximum of its `CDR`. So, we might guess that a list containing two numbers, where the first is bigger than the second (for example, `10` followed by `3`), would provide a counterexample. Let us use the execution facility offered by Nqthm, by issuing the command `(R-LOOP)`, to verify that such a list provides a counterexample to the goal above. In the following display, terms following the prompt '`*`' are the input, and as always, semicolons precede comments.

```
*(let ((x '(10 3))) ; Bind x to the list suggested above.
      (implies
       (and (listp x)
            (not (equal (maxlist (cdr x)) (car x)))
            (not (lessp (maxlist (cdr x))
                        (maxlist (delete1 (maxlist (cdr x))
                                          (cdr x)))))
            (member (maxlist (cdr x)) (cdr x))
            (not (lessp (car x) (maxlist (cdr x)))))
       (not (numberp (car x)))))
 F ; Aha!  The goal is false using the indicated value of x.
*(let ((x '(10 3)) (b 3))
   ; Let us see if we can construct a counterexample to the theorem
   ; maxlist-delete1-rearrange, which we are in the process of
   ; trying to prove.
      (implies (member b x)
               (equal (maxlist (cons b (delete1 b x)))
                      (maxlist x))))
 F ; Aha!  The ''theorem'' is false using the indicated values of x
   ; and b.
*(let ((x '(10 3)) (b 3))
   ; Let us see what the values of some of the terms are in the
   ; counterexample above.
   (list (maxlist (cons b (delete1 b x)))
         (delete1 b x)
         (maxlist x)))
 '(3 NIL 10)
```

Aha! We find that `(DELETE1 B X)` returns `NIL` in our example, even though for the indicated values of `B` and `X` we would expect the result to be a list containing `10`. And sure enough, when we look back at the definition of `DELETE1`, we find a bug: when we recur, we need to return

```
(cons (car x) (delete1 elt (cdr x))),
```

not

```
(delete1 elt (cdr x)).
```

So, we undo the definition of DELETE1:

```
(ubt)
```

(When UBT is called with no argument, it removes only the most recent event from the current history. If we had discovered much later in the course of our proof effort that DELETE1 contained a bug, i.e., if we had introduced other events after the definition of DELETE1, we should have typed the command (UBT DELETE1); see Subsection 2.3. Of course, this has the 'drawback' that all the events based on the buggy definition have to be processed again, but in this way we are assured that the (logical) consistency of the current history is maintained.)

Here is the correct definition for DELETE1:

```
(DEFN DELETE1 (ELT X)
  (IF (LISTP X)
      (IF (EQUAL ELT (CAR X))
          (CDR X)
        (CONS (CAR X) (DELETE1 ELT (CDR X))))
    X))
```

And now, the proof of MAXLIST-DELETE1-REARRANGE succeeds.

5.5. Outline of Termination Proof

Our initial attempts to admit the definition of INDUCTION-FN, shown in the preceding subsection, are serious failures. The main discovery during that process is a reaffirmation of an important idea when using Nqthm: One must try to keep track of where he/she is going, formulating lemmas that will lead to the goal at hand. In particular, we have found that if we submit the definition by using VERIFY-DEFN in place of DEFN, the Pc-Nqthm interactive loop presents us with two goals, one of which is easy for Nqthm to simplify to T. Let us state the other as an axiom, for now, following the same general approach elucidated earlier in Subsection 5.2: top-down proof. To increase readability we reintroduce the variable X0 to represent the list constructed in the induction scheme.

```
(add-axiom induction-fn-help-2 (rewrite)
  (let ((x0 (cons (sub1 (maxlist x))
                  (cons (add1 (minlist x))
                        (delete1 (maxlist x)
                                 (delete1 (minlist x) x))))))
    (implies
```

```
(and (equal (sumlist x) (times k (length x)))
     (not (equal (minlist x) (maxlist x))))
(ord-lessp
 (cons (add1 (maxlist x0)) (occurrences (maxlist x0) x0))
 (cons (add1 (maxlist x)) (occurrences (maxlist x) x))))))
```

We formulate this axiom so that Nqthm can immediately admit the definition of INDUCTION-FN. In order to make that happen, we would like to instruct Nqthm that it should ignore the definitions of all the functions that appear in the definition of INDUCTION-FN or in the axiom above. We do so because we believe that we have already proved (or assumed) exactly what we need, and if Nqthm applies any additional information to the problem, the direction of the proof may somehow change, eliminating the applicability of rewrite rules (especially, of the axiom introduced above). Fortunately, Nqthm supports the capability to control which rules (especially, which definitions and which rewrite rules) are 'enabled', i.e., are allowed to be used automatically by the theorem prover. Rules (including definitions) are enabled by default, but the user may disable them.

We may disable the relevant functions individually. However, it is convenient to assign a name, say INDUCTION-FN-DISABLES, to the appropriate list of function names.

```
(deftheory induction-fn-disables
 (sumlist times length minlist maxlist delete1 occurrences))
```

Then, we can *disable* all those names at once.

```
(disable-theory induction-fn-disables)
```

At this point Nqthm accepts the definition of INDUCTION-FN. For future reference, we note that ENABLE-THEORY is the analogous command for telling Nqthm to 'remember again' the rules in the specified theory.

We turn now to the task of proving the ADD-AXIOM formula above. For, once we replace that event by a corresponding PROVE-LEMMA event (and make sure that we have also executed the DISABLE-THEORY event above), then Nqthm should (we hope) again accept the definition of INDUCTION-FN. At this point we could use our induction scheme to begin the proof of the 'main lemma', deferring until later the proof of the ADD-AXIOM event above. However, our preference, though not a particularly strong one, is to proceed as already indicated: back up and prove the relevant axiom. That way we do not leave this loose end to deal with later. So, we undo all events back through the axiom INDUCTION-FN-HELP-2, using the following command.

```
(ubt induction-fn-help-2)
```

Our first step is to break the axiom above into two cases, corresponding to Cases 2.2 and 2.1 in the hand proof in Lemma 2, i.e., corresponding to whether or not the maximum occurs twice in the list. Continuing in a top-down decompositional style, we formulate the two cases now as axioms. We will use these axioms explicitly (see the discussion of USE hints, Subsection 5.2), so we do not label them for automatic use as rewrite rules.

```
(add-axiom induction-fn-help-2-max-occurs-twice ()
  ; Case 2.2 from the hand proof.
  (let ((x0 (cons (sub1 (maxlist x))
                  (cons (add1 (minlist x))
                        (delete1 (maxlist x)
                                 (delete1 (minlist x) x))))))
    (implies
     (and (equal (sumlist x) (times k (length x)))
          (not (equal (minlist x) (maxlist x)))
          (lessp 1 (occurrences (maxlist x) x)))
     (ord-lessp
      (cons (add1 (maxlist x0)) (occurrences (maxlist x0) x0))
      (cons (add1 (maxlist x)) (occurrences (maxlist x) x))))))

(add-axiom induction-fn-help-2-max-occurs-once ()
  ; Case 2.1 from the hand proof.
  (let ((x0 (cons (sub1 (maxlist x))
                  (cons (add1 (minlist x))
                        (delete1 (maxlist x)
                                 (delete1 (minlist x) x))))))
    (implies
     (and (equal (sumlist x) (times k (length x)))
          (not (equal (minlist x) (maxlist x)))
          (not (lessp 1 (occurrences (maxlist x) x))))
     (ord-lessp
      (cons (add1 (maxlist x0)) (occurrences (maxlist x0) x0))
      (cons (add1 (maxlist x)) (occurrences (maxlist x) x))))))
```

And now we may make the desired conclusion by instructing Nqthm to use the above axioms explicitly. Since we expect the following theorem to follow trivially from the two cases, we instruct the theorem prover to disable all rules in the database, using the (DISABLE-THEORY T) hint shown below, except for those rules (such as the definition of IMPLIES) that are built in, by using the (ENABLE-THEORY GROUND-ZERO) hint.

```
(prove-lemma induction-fn-help-2 (rewrite)
  (let ((x0 (cons (sub1 (maxlist x))
                  (cons (add1 (minlist x))
                        (delete1 (maxlist x)
                                 (delete1 (minlist x) x))))))
    (implies
     (and (equal (sumlist x) (times k (length x)))
          (not (equal (minlist x) (maxlist x))))
     (ord-lessp
      (cons (add1 (maxlist x0)) (occurrences (maxlist x0) x0))
      (cons (add1 (maxlist x)) (occurrences (maxlist x) x)))))
  ((use (induction-fn-help-2-max-occurs-twice)
        (induction-fn-help-2-max-occurs-once))
```

```
   (disable-theory t)
   (enable-theory ground-zero)))
```

Finally, let us undo, executing the command

```
(ubt induction-fn-help-2-max-occurs-twice)
```

in anticipation of proving the two cases. Now that we are ready to proceed in a 'forward' manner, we submit the DEFTHEORY event that we 'undid' earlier.

5.6. Proof of Lemma 1 from 'Hand Proof'

Let us prove a formalization of Lemma 1 from the hand proof in Section 4: if the maximum and minimum of a list are distinct, then the maximum is at least 2 greater than the minimum. We are starting the proof of Lemma 1 simply because it seems fun at the moment, and because we have already seen in our hand proof that we'll need this fact. It is often important when using Nqthm not to waste time proving irrelevant facts merely because they *seem* relevant. But, we are confident that we will indeed need the following fact (which we will call MINLIST-LESS-THAN-MAXLIST-MINUS-1).

```
(implies (and (listp x)
              (equal (sumlist x) (times k (length x)))
              (not (equal (minlist x) (maxlist x))))
         (lessp (minlist x) (sub1 (maxlist x))))
```

Although this fact seems obvious, we do not immediately see an obvious proof of this theorem, so we do not expect Nqthm to see a proof either, without some help. What is our plan? We follow the outline presented in the hand proof, Subsection 5.1. We consider the following two events, which turn out to be useful later.

```
(PROVE-LEMMA MAX-GREATER-THAN-AVERAGE (REWRITE)
             (IMPLIES (AND (LISTP X)
                           (GEQ (SUMLIST X) (TIMES K (LENGTH X)))
                           (NOT (EQUAL (FIX K) (MAXLIST X))))
                      (LESSP K (MAXLIST X))))

(PROVE-LEMMA MIN-LESS-THAN-AVERAGE (REWRITE)
             (IMPLIES (AND (LISTP X)
                           (LEQ (SUMLIST X) (TIMES K (LENGTH X)))
                           (NOT (EQUAL (FIX K) (MINLIST X))))
                      (LESSP (MINLIST X) K)))
```

Clearly their conclusions together show that (MINLIST X) + 1 < (MAXLIST X). However, we confess that probably our thinking is not clear as we formulate these lemmas; it's not obvious just how they help us. Anyhow, among our thoughts are the following, which illustrate some generally useful ideas.

1. Each of these lemmas involves only one function from {MINLIST, MAXLIST}. Often it is easier to reason about fewer functions at any given time.

2. The equality hypothesis in our main goal does not seem amenable to proof by induction, because it is not preserved when we replace X with (CDR X). To be honest: rather than thinking this point through carefully, instead we just press on, hoping that our intuition is correct.

At this point we wonder if the lemmas above are enough to prove MINLIST-LESS-THAN-MAXLIST-MINUS-1. We try it, but the proof fails. That is not surprising, because probably first we have to use the lemmas above to show that if (MAXLIST X) $\neq$ (MINLIST X), then neither one is equal to K. At any rate, after some fumbling around, we decide to prove the following lemma (that we are able to prove only after proving a lemma stating that (TIMES (LENGTH X) (MINLIST X)) is not greater than (SUMLIST X)):

```
(PROVE-LEMMA
    MINLIST-NOT-MAXLIST-IMPLIES-MINLIST-LESSP-AVERAGE-LEMMA
    ()
    (IMPLIES (NOT (EQUAL (MINLIST X) (MAXLIST X)))
             (LESSP (TIMES (MINLIST X) (LENGTH X))
                    (SUMLIST X))))
```

Notice that we have not made it a rewrite rule. This is because we will be USE-ing it to prove the lemma below. If one gives a rewrite rule, say R, as a USE hint, then one is proving (IMPLIES R...), and there is a danger that R will rewrite to T, thus making R USEless!

Then, the following is an immediate consequence of the lemma above, by plugging in the equality hypothesis below into the lemma above. It does not prove automatically, so we'll just give a USE hint. The use of FIX here is not important.

```
(PROVE-LEMMA MINLIST-NOT-MAXLIST-IMPLIES-MINLIST-LESSP-AVERAGE
 (REWRITE)
 (IMPLIES (AND (LISTP X)
               (EQUAL (SUMLIST X) (TIMES K (LENGTH X)))
               (NOT (EQUAL (MINLIST X) (MAXLIST X))))
          (LESSP (MINLIST X) (FIX K)))
 ((USE (MINLIST-NOT-MAXLIST-IMPLIES-MINLIST-LESSP-AVERAGE-LEMMA))))
```

Now we try to prove the analog of the lemma MINLIST-NOT-MAXLIST-IMPLIES-MINLIST-LESSP-AVERAGE-LEMMA, modified for MAXLIST in place of MINLIST. The proof fails, but it succeeds after we have proved an analog of the lemma we needed to prove MINLIST-NOT-MAXLIST-IMPLIES-MINLIST-LESSP-AVERAGE-LEMMA.

The main sublemma follows by the analogous argument:

```
(PROVE-LEMMA MINLIST-NOT-MAXLIST-IMPLIES-MAXLIST-GREATERP-AVERAGE
  (REWRITE)
```

```
  (IMPLIES (AND (LISTP X)
                (EQUAL (SUMLIST X) (TIMES K (LENGTH X)))
                (NOT (EQUAL (MINLIST X) (MAXLIST X))))
           (LESSP (FIX K) (MAXLIST X)))
  ((USE
    (MINLIST-NOT-MAXLIST-IMPLIES-MAXLIST-GREATERP-AVERAGE-LEMMA
     ))))
```

In fact, the lemma below now proves! But, it requires the USE hints that are given to it. But then, why did not we disable the lemma above (see the comment below MINLIST-NOT-MAXLIST-IMPLIES-MINLIST-LESSP-AVERAGE-LEMMA), since we generally try to disable rewrite rules that are given to USE hints? As explained above, the issue is that we do not want a USEd hypothesis to rewrite to T. However, the theorem prover reports that the rule above is a *linear rule*; actually, as explained in [6], the theorem prover creates a linear rule when the top-level function symbol in the conclusion is LESSP, or even when the conclusion is the negation of such a term. Linear rules are used only to derive contradictions, not to rewrite, and hence the concern about USE hints is groundless in this case.

```
(PROVE-LEMMA
    MINLIST-LESS-THAN-MAXLIST-MINUS-1 (REWRITE)
    (IMPLIES (AND (LISTP X)
                  (EQUAL (SUMLIST X) (TIMES K (LENGTH X)))
                  (NOT (EQUAL (MINLIST X) (MAXLIST X))))
             (LESSP (MINLIST X) (SUB1 (MAXLIST X))))
    ((USE
      (MINLIST-NOT-MAXLIST-IMPLIES-MINLIST-LESSP-AVERAGE)
      (MINLIST-NOT-MAXLIST-IMPLIES-MAXLIST-GREATERP-AVERAGE))))
```

5.7. Proof of Hand Proof's Case 2.2

Our goal in this section is to prove the lemma INDUCTION-FN-HELP-2-MAX-OCCURS-TWICE, stated in Subsection 5.5. Informally put, this lemma says that if the maximum of X occurs at least twice in X, then X0 (defined earlier, Subsection 5.3) is 'smaller' than X in an appropriate sense. This lemma follows immediately from the following two claims, which say that when X is replaced by X0, then the maximum stays the same, but it occurs fewer times. In fact, after executing these two ADD-AXIOM events, we are able to prove the lemma, with the hint (disable-theory induction-fn-disables):

```
(add-axiom induction-fn-help-2-max-occurs-twice-lemma-1 (rewrite)
  (let ((x0 (cons (sub1 (maxlist x))
                  (cons (add1 (minlist x))
                        (delete1 (maxlist x)
                                 (delete1 (minlist x) x))))))
    (implies
     (and (equal (sumlist x) (times k (length x)))
```

```
          (not (equal (minlist x) (maxlist x)))
          (lessp 1 (occurrences (maxlist x) x)))
     (equal (maxlist x0)
            (maxlist x)))))

(add-axiom induction-fn-help-2-max-occurs-twice-lemma-2 (rewrite)
  (let ((x0 (cons (sub1 (maxlist x))
                  (cons (add1 (minlist x))
                        (delete1 (maxlist x)
                                 (delete1 (minlist x) x))))))
    (implies
     (and (equal (sumlist x) (times k (length x)))
          (not (equal (minlist x) (maxlist x)))
          (lessp 1 (occurrences (maxlist x) x)))
     (equal (lessp (occurrences (maxlist x)
                                x0)
                   (occurrences (maxlist x) x))
            t))))
```

Notice that the conclusion of the second lemma above has the form `(EQUAL ... T)`. When we did not use this form, the system complained on an earlier version of this lemma (which had a 'typo') that it was trying to store this as a linear rule, and reported that it could not do so. Rather than think about this problem, we simply wrote the conclusion as an `EQUAL` term rather than a `LESSP` term, so that Nqthm will store it as an ordinary replacement rule. In fact, the present version can actually be stored as a linear rule, but it seems reasonable to take the simpler route of just using rewriting, which is somehow easier to understand.

The Nqthm rewriter works from the inside out; it rewrites all arguments of a term before rewriting the entire term. Therefore, we apply `MAXLIST` to `X`, not to `X0`, in the second lemma. For if the second lemma's left side (the `LESSP` term in the equality) contained an occurrence of `(MAXLIST X0)`, it would probably not match any term encountered by the rewriter, because the first lemma will already have been used to rewrite `(MAXLIST X0)` to `(MAXLIST X)`.

After undoing these two axioms, we turn to the task of proving the first of them. We think that the theorem is too hard to prove automatically, and so we decide to prove some lemmas to simplify the problem. In particular, we wish to prove rewrite rules that will simplify the term

```
(maxlist (delete1 (maxlist x) (delete1 (minlist x) x))),
```

since we expect this term to appear after the simplifier applies the definition of `MAXLIST` twice. (Occasionally it is useful to override our general reluctance to prove lemmas that *might* be needed, when we can see that they probably *will* be needed in order to simplify terms.)

Rather than try to reason directly about this term, we formulate two simple lemmas. The first implies that neither call of `DELETE1` in this term changes the maximum, except that the second lemma is necessary in order to guarantee that the maximum still occurs twice after deletion of the minimum.

```
(PROVE-LEMMA
 DELETE1-PRESERVES-MAXLIST-WHEN-MAXLIST-OCCURS-MORE-THAN-ONCE
 (REWRITE)
 (IMPLIES (LESSP 1 (OCCURRENCES (MAXLIST X) X))
          (EQUAL (MAXLIST (DELETE1 A X)) (MAXLIST X))))

(PROVE-LEMMA DELETE1-OCCURRENCES (REWRITE)
             (EQUAL (OCCURRENCES A (DELETE1 B X))
                    (IF (AND (EQUAL A B)
                             (MEMBER B X))
                        (SUB1 (OCCURRENCES A X))
                      (OCCURRENCES A X))))
```

Now with some additional work, we may prove INDUCTION-FN-HELP-2-MAX-OCCURS-TWICE-LEMMA-1, INDUCTION-FN-HELP-2-MAX-OCCURS-TWICE-LEMMA-2, and finally our goal INDUCTION-FN-HELP-2-MAX-OCCURS-TWICE (see Subsection 5.5). See [21] for details.

5.8. Proof of Hand Proof's Case 2.1

An analysis similar to that at the start of the preceding subsection suggests that the following goal suffices for proving Case 2.1 from the hand proof. In fact our approach is less thoughtful; for details see [21].

```
(add-axiom induction-fn-help-2-max-occurs-once-main-lemma (rewrite)
  (implies
   (and
    (equal (sumlist x) (times k (length x)))
    (not (equal (minlist x) (maxlist x)))
    (not (lessp 1 (occurrences (maxlist x) x))))
   (equal (lessp (maxlist (cons (sub1 (maxlist x))
                                (cons (add1 (minlist x))
                                      (delete1 (maxlist x)
                                               (delete1 (minlist x)
                                                        x))))))
                 (maxlist x))
          t)))
```

In fact, this axiom suffices to prove

INDUCTION-FN-HELP-2-MAX-OCCURS-ONCE.

The main obstacle to proving this lemma (axiom) is that our formulation of

MINLIST-LESS-THAN-MAXLIST-MINUS-1

(see Subsection 5.6) is too weak. In fact, its first hypothesis, (LISTP X), follows from its last, (NOT (EQUAL (MINLIST X) (MAXLIST X))). By dropping its first hypothesis we can prove a better version that helps us finish the proof of the axiom above. For details see [21].

5.9. Completion of Termination Proof

Finally, we may prove INDUCTION-FN-HELP-2 and complete the proof of termination. All we have to do is submit the definition whose termination we have been proving, surrounding it with the theory management that we introduced earlier in this section. Thus, the first event below instructs the theorem prover to ignore the rules in the theory INDUCTION-FN-DISABLES (which was earlier introduced by a DEFTHEORY event), while the final event returns those rules to their enabled status.

```
(DISABLE-THEORY INDUCTION-FN-DISABLES)

(DEFN INDUCTION-FN (X K)
  (IF (NOT (EQUAL (SUMLIST X) (TIMES K (LENGTH X))))
      T
    (IF (EQUAL (MINLIST X) (MAXLIST X))
        T
      (INDUCTION-FN (CONS (SUB1 (MAXLIST X))
                          (CONS (ADD1 (MINLIST X))
                                (DELETE1 (MAXLIST X)
                                         (DELETE1 (MINLIST X) X))))
                    K)))
  ((ORD-LESSP (CONS (ADD1 (MAXLIST X)) (OCCURRENCES (MAXLIST X)
                                                    X)))))

(ENABLE-THEORY INDUCTION-FN-DISABLES)
```

5.10. The Rest of the Proof

Now we attack our main lemma. The hint (INDUCT (INDUCTION-FN X K)) shown below tells the prover to use the induction scheme suggested by the term (INDUCTION-FN X K), in the sense described in Subsections 5.3 and (more generally) 2.4.

```
(prove-lemma main-lemma ()
             (let ((n (length x)))
               (implies (equal (sumlist x) (times k n))
                        (not (lessp (exp k n) (prodlist x)))))
             ((induct (induction-fn x k))))
```

The proof fails, but that is not surprising: so far we have concentrated only on the proof of termination of INDUCTION-FN. (Admittedly, that is probably by far the hardest part; but, it's not the only part.) We inspect the proof, and this suggests that we prove a simple lemma about terms of the form (SUMLIST (DELETE1 ...)).

At this point the proof of the main lemma still fails, and we wake up and say to ourselves: "We haven't even proved the base case yet!" Why should we perhaps do the base case first?

1. It seems reasonable to prove easy things first, because sometimes one develops useful rewrite rules in the process that help simplify the harder goals into more comprehensible forms.

2. Generally speaking, base cases tend to be easier than inductive steps.

Below, we show the lemma proved for the base case.

```
(PROVE-LEMMA MAIN-LEMMA-BASE-CASE (REWRITE)
             (IMPLIES (AND (EQUAL (SUMLIST X) (TIMES K (LENGTH X)))
                           (EQUAL (MINLIST X) (MAXLIST X)))
                      (EQUAL (EXP K (LENGTH X)) (PRODLIST X)))
             ((USE (MAIN-LEMMA-BASE-CASE-LEMMA-1)
                   (MAIN-LEMMA-BASE-CASE-LEMMA-2))))
```

For the sake of completeness we show also the facts stated by the two lemmas in the USE hint above.

```
(implies (equal (minlist x) (maxlist x))
         (equal (exp (minlist x) (length x)) (prodlist x)))

(implies (and (listp x)
              (equal (sumlist x) (times k (length x)))
              (equal (minlist x) (maxlist x)))
         (equal (minlist x) (fix k)))
```

We turn now to the inductive step. Since we just returned from a weekend break, we get our bearing by entering the Pc-Nqthm interactive loop. Briefly put, we are led to the following lemma: `(IMPLIES (LISTP X) (MEMBER (MINLIST X) X))`. This is clearly not always true, for example if X contains a single non-numeric element (in which case `(MINLIST X)` is 0, which does not belong to X).

To fix this problem, let us work in a somewhat more 'typed' framework by assuming in our top-level theorem that we have a list of numbers. So, at this point we redo all of our work, but add the following definition:

```
(DEFN NUMBERP-LISTP (X)
  (IF (LISTP X)
      (AND (NUMBERP (CAR X))
           (NUMBERP-LISTP (CDR X)))
    (EQUAL X NIL)))
```

We'll try to fix things by adding `(NUMBERP-LISTP X)` as a top-level hypothesis when we finally glue all this together.

And now, an adjusted ('typed') version of the lemma mentioned above can be proved.

```
(PROVE-LEMMA MEMBER-MINLIST (REWRITE)
             (IMPLIES (AND (LISTP X)
                           (NUMBERP-LISTP X))
                      (MEMBER (MINLIST X) X)))
```

Here is the revised version of the main lemma. The DO-NOT-INDUCT hint below is simply a device to get the theorem prover to abort the proof rather than push a goal to be proved by a (sub-)induction.

```
(prove-lemma main-lemma ()
             (let ((n (length x)))
               (implies (and (numberp-listp x)
                             (equal (sumlist x) (times k n)))
                        (not (lessp (exp k n) (prodlist x)))))
             ((induct (induction-fn x k))
              (do-not-induct t)))
```

The main lemma fails. A long interaction with the prover leads us to the heart of the arithmetic argument, which corresponds in the hand proof to the reasoning that allows us to prove Equation (1) in Section 4. We can represent this argument informally on paper as follows, where we write the product of the original list as `min * max * rest`.

```
(min + 1) * (max - 1) * rest
=
[(min * max) + max - min - 1] * rest
=
(min * max * rest) + ([max - min - 1] * rest)
=
prodlist(x) + ([max - min - 1] * rest)
```

As usual, we would like to state facts in terms of useful rewrite rules when possible. When that seems too awkward or difficult, we can try to formulate lemmas that can be used, with USE hints, to give the desired result. However, let us see if the direct approach of proving a single rewrite rule that rewrites an appropriate term can be used. The following lemma formalizes the content of the informal equations displayed above.

```
(PROVE-LEMMA
    PRODUCT-OF-MODIFIED-LIST (REWRITE)
    (IMPLIES (AND (LESSP MIN MAX)
                  (MEMBER MIN X)
                  (MEMBER MAX X))
             (LET ((REST (PRODLIST (DELETE1 MAX
                                            (DELETE1 MIN X)))))
                  (EQUAL (TIMES (ADD1 MIN) (SUB1 MAX) REST)
                         ; prodlist(x) + ([max - min - 1] * rest)
                         (PLUS (PRODLIST X)
                               (TIMES (DIFFERENCE MAX (ADD1 MIN))
                                                        REST)))))
    ((USE (POSITIVE-IMPLIES-NUMBERP))))
```

The lemma POSITIVE-IMPLIES-NUMBERP in the USE hint above states that if (LESSP MIN MAX), then MAX is a number. We had to prove this lemma because the linear arithmetic decider was not able to infer it automatically in the course of the proof of PRODUCT-OF-MODIFIED-LIST.

And now the main lemma goes through!

```
(PROVE-LEMMA MAIN-LEMMA ()
             (LET ((N (LENGTH X)))
               (IMPLIES (AND (NUMBERP-LISTP X)
                             (EQUAL (SUMLIST X) (TIMES K N)))
                        (NOT (LESSP (EXP K N) (PRODLIST X)))))
             ((INDUCT (INDUCTION-FN X K))
              (DO-NOT-INDUCT T)))
```

Then we paste down the events referred to in Subsection 5.2 to prove the theorem from MAIN-LEMMA.

Note, however, that since we have introduced consideration of NUMBERP-LISTP into the problem, a little further work may be required. And indeed, when we try our main theorem, the proof fails. However, the checkpoint tool suggests to us the lemma (see below) that we need to prove the main theorem!

```
(PROVE-LEMMA NUMBERP-LISTP-SCALAR-PRODUCT (REWRITE)
             (IMPLIES (NUMBERP-LISTP A)
                      (NUMBERP-LISTP (SCALAR-PRODUCT N A))))

(PROVE-LEMMA MAIN ()
             (IMPLIES (NUMBERP-LISTP A)
                      (LET ((N (LENGTH A)))
                        (GEQ (EXP (SUMLIST A) N)
                             (TIMES (EXP N N) (PRODLIST A)))))
             ((USE (MAIN-LEMMA (K (SUMLIST A))
                               (X (SCALAR-PRODUCT
                                   (LENGTH A) A))))))
```

5.11. Refining the Theorem

Let us introduce a version of this theorem that more closely resembles the informal statement of Section 4, by using the iteration construct FOR provided by Nqthm. We do not explain FOR here, except to say that it's similar to looping constructs in Lisp and other languages and is documented, of course, in [6]; we believe that its meaning is reasonably self-evident. Our approach here is simply to note that the new version of MAIN is an immediate consequence of MAIN once we tell the rewriter how to write SUMLIST and PRODLIST in terms of FOR.

```
(PROVE-LEMMA SUMLIST-FOR (REWRITE)
             (EQUAL (SUMLIST A)
                    (FOR X IN A SUM X)))

(PROVE-LEMMA PRODLIST-FOR (REWRITE)
             (EQUAL (PRODLIST A)
                    (FOR X IN A MULTIPLY X)))

(PROVE-LEMMA MAIN-AGAIN ()
             (IMPLIES (AND (NUMBERP-LISTP A)
```

```
                (GEQ (EXP (FOR X IN A SUM X) N)
                     (TIMES (EXP N N)
                            (FOR X IN A MULTIPLY X))))
        ((USE (MAIN))))
```

6. Conclusion

We have illustrated the following two-part approach for using Nqthm (as explained in Section 3). This is a bit of a simplification of our approach, but it is a good first approximation.

- Begin with a high level strategy, decomposing the entire theorem into lemmas, based on the structure of a hand proof.
- Submit lemmas to the theorem prover in support of the high-level strategy.

A main point of this paper is that when interacting with the Nqthm prover (the second part of the approach above), it is *much* less important to be able to *predict* what the theorem prover will do than it is to learn how to *react* to what the theorem prover *does*. It is generally useful to view Nqthm as a black box that either succeeds in its proof attempt or else generates an unproved, simplified goal. Such a goal will often suggest the opportunity to formulate a rewrite rule that would cause the prover to create a simpler, perhaps even proved, version of that goal on subsequent proof attempts.

We will resist the (not very strong) temptation to try to make in-depth comparisons between the hand proof and the mechanized proof. For one thing, the hand proof is lacking in many details. In a certain sense, actually, the mechanized proof is lacking in many details too; after all, the theorem prover has some built-in knowledge, and we start our proof by including an existing library of arithmetic definitions and proved lemmas. Nevertheless, we can give some raw numbers. The LaTeX source for the hand-proof section contains approximately 4.5K characters, while the replayable Nqthm events comprise 19.4K characters. The Nqthm proof has taken just under two minutes to replay on a Sparc 20.

A list of helpful tips for using Nqthm may be found in [21].

Notes

[1] Named after a directory on which the system was originally developed.

[2] This system, and its Pc-Nqthm interactive enhancement (described below), are currently available from Internet host ftp.cli.com: directory /pub/nqthm/nqthm-1992/ for the Nqthm-1992 release, and directory /pub/pc-nqthm/pc-nqthm-1992/ for the Pc-Nqthm-1992 release.

[3] See, however, any of [15, 16, 1, 17] for introductions to the use of Nqthm. In fact [17] gives a detailed account of the use of Nqthm and its interactive enhancement to do a non-trivial proof. However, it has a different emphasis from the present paper's tutorial direction; for example, it omits the hints given to Nqthm, which are an important part of the process of interacting with the system. There are other introductions to Nqthm. For example, John Cowles has written a brief introduction in the file examples/cowles/intro-eg.events contained in the Nqthm-1992 release. The paper [10] provides an introduction to Nqthm as part of a verification system.

[4] The initials 'Pc' stand for 'Proof-checker', an odd reference to the lower-level nature of the interaction made possible in this system.

[5] Unlike some other Lisp dialects, the predicate `LISTP` in the Nqthm object recognizes ordered pairs and *nonempty* lists, *not* the empty list.

[6] The tool is available with the Pc-Nqthm-1992 distribution; see misc/checkpoints.doc inside that distribution.

[7] A referee expressed the reasonable concern that one may need knowledge of how the system works in order to be able to inspect the rewrite stack. In our view, it is fair to say that this only requires user-level knowledge, not knowledge of the details of the underlying implementation.

[8] Fausto Giunchiglia has made this observation to us. He also pointed out that the central role of interaction in theorem proving was initially advocated by McCarthy and Weyhrauch.

[9] In fact, this library is available in the Nqthm-1992 release, in the file examples/numbers/naturals.events, which has many contributors, including William Bevier, Bob Boyer, Matt Kaufmann, J Moore, and Matt Wilding.

[10] Actually `n` is optional; when it is omitted, all names from the history are printed out.

[11] Actually this order is represented in the Nqthm logic by `ORD-LESSP`, an order on notations for ordinals less than ϵ_0; see [6].

[12] In our particular environment, this amounts to issuing two `control-c` characters.

[13] The 'moment' is actually the next day. After logging back in, we used the Nqthm utility `DO-FILE` to process the file of events created so far. A benefit of such breaks is that `DO-FILE` will fail if we have not put the appropriate events in the file we are constructing, and we prefer to find that out now and fix it, rather than waiting until the very end of the proof!

[14] One referee pointed out that the following subgoal is easier to understand if one realizes that the induction scheme chosen is based on the definition of `MAXLIST`. While this is true in a sense, we are trying to illustrate an approach that emphasizes inspection of a 'checkpoint' goal, and we believe that it is often distracting to the user to consider how that goal arises unless direct inspection fails to suggest a course of action.

[15] CLI technical reports are available from Web server http://www.cli.com. (See refs.)

[16] IRST technical reports are available from Web server http://afrodite.itc.it:1024. (See refs.)

References

1. Basin, D. and Kaufmann, M.: The Boyer–Moore prover and Nuprl: An experimental comparison, in *Proc. Workshop for Basic Research Action, Logical Frameworks*, Antibes, France, May 1990.
2. Bevier, W., Hunt, W., Jr., Moore, J, and Young, W.: An approach to systems verification, *J. Automated Reasoning* **5** (1989), 411–428.
3. Boyer, R. S. and Moore, J S.: *A Computational Logic*, Academic Press, 1979.
4. Borrione, D., Pierre, L., Salem, A., and Ashraf, M.: Formal verification of VHDL descriptions in the PREVAIL environment, in *IEEE Design and Test*, June 1992.
5. Boyer, R. S. and Moore, J S.: Metafunctions: proving them correct and using them efficiently as new proof procedures, in R. S. Boyer and J S. Moore (eds), *The Correctness Problem in Computer Science*, Academic Press, London, 1981.
6. Boyer, R. S. and Moore, J S.: *A Computational Logic Handbook*, Academic Press, 1988.
7. Boyer, R. S., Kaufmann, M., and Moore, J S.: The Boyer–Moore theorem prover and its interactive enhancement, *Computers and Mathematics with Applications* **29** (1995), 27–62.
8. Bundy, A.: Talk in Challenge Problems section of Workshop on the Automation of Proof by Mathematical Induction, co-sponsored by MInd and IndUS, July 11–12, 1993; at AAAI-93 11th *National Conf. Artificial Intelligence*, Washington DC, USA.
9. Good, D. and Young, W.: Mathematical methods for digital systems development, in S. Prehn and W. J. Toetenel (eds), *VDM'91 Formal Software Development Methods*, Springer-Verlag *Lecture Notes in Computer Science* 552 (1991), 406–430.

10. Johnson, S. and Nagle, J.: *Pascal-F Verifier User's Manual, Version 2*, Ford Aerospace & Communications Corporation, Palo Alto, CA, 1986.
11. Kaufmann, M.: A User's Manual for an Interactive Enhancement to the Boyer–Moore Theorem Prover, Technical Report 19, Computational Logic, Inc., May 1988.[15]
12. Kaufmann, M.: Addition of Free Variables to the PC-NQTHM Interactive Enhancement of the Boyer–Moore Theorem Prover, Technical Report 42, Computational Logic, Inc., March 1990.[15]
13. Kaufmann, M.: Response to FM91 Survey of Formal Methods: Nqthm and Pc-Nqthm, Technical Report 75, Computational Logic, Inc., March 1992.[15]
14. Kaufmann, M.: An Assistant for Reading Nqthm Proof Output, Technical Report 85, Computational Logic, Inc., November 1992.[15]
15. Kaufmann, M.: An example in NQTHM: Ramsey's theorem, Internal Note 100, Computational Logic, Inc., November 1988.
16. Kaufmann, M.: An instructive example for beginning users of the Boyer–Moore theorem prover, Internal Note 185, Computational Logic, Inc., April 1990.
17. Kaufmann, M.: Generalization in the presence of free variables: A mechanically-checked correctness proof for one algorithm, *J. Automated Reasoning* **7** (1991), 109–158.
18. Kaufmann, M.: An extension of the Boyer–Moore theorem prover to support first-order quantification, *J. Automated Reasoning* **9** (1992), 355–372.
19. Pierre, L.: The formal proof of sequential circuits described in CASCADE using the Boyer–Moore theorem prover, in L. Claesen (ed.), *Formal VLSI Correctness Verification*, North-Holland, 1990.
20. Stallman, R. M.: *GNU EMACS Manual, 6th edn.*, Free Software Foundation, March 1987.
21. Kaufmann, M. and Pecchiari, P.: Interaction with the Boyer–Moore Theorem Prover: A Tutorial Study Using the Arithmetic–Geometric Mean Theorem, Technical Report 100, Computational Logic, Inc., August 1994,[15] and Technical Report 9409-01, IRST, August 1994.[16] (Revised June 1995.)

Printed in the United States
71954LV00002B/95

9 780792 340102